Thomas Jeferson Vieira
Luciano Fleischfresser
Lucas Augusto Prudente

How do green roofs change the thermal balance indoors?

AF524487

Thomas Jeferson Vieira
Luciano Fleischfresser
Lucas Augusto Prudente

How do green roofs change the thermal balance indoors?

Comparison of internal temperature variation in two rooms clad with different roofs

ScienciaScripts

Imprint

Any brand names and product names mentioned in this book are subject to trademark, brand or patent protection and are trademarks or registered trademarks of their respective holders. The use of brand names, product names, common names, trade names, product descriptions etc. even without a particular marking in this work is in no way to be construed to mean that such names may be regarded as unrestricted in respect of trademark and brand protection legislation and could thus be used by anyone.

Cover image: www.ingimage.com

This book is a translation from the original published under ISBN 978-3-330-99563-5.

Publisher:
Sciencia Scripts
is a trademark of
Dodo Books Indian Ocean Ltd. and OmniScriptum S.R.L publishing group

120 High Road, East Finchley, London, N2 9ED, United Kingdom
Str. Armeneasca 28/1, office 1, Chisinau MD-2012, Republic of Moldova, Europe
Managing Directors: Ieva Konstantinova, Victoria Ursu
info@omniscriptum.com

Printed at: see last page
ISBN: 978-620-8-58184-8

Copyright © Thomas Jeferson Vieira, Luciano Fleischfresser, Lucas Augusto Prudente
Copyright © 2025 Dodo Books Indian Ocean Ltd. and OmniScriptum S.R.L publishing group

ACKNOWLEDGEMENTS

Firstly, I would like to thank my mother Neusa Moreira Vieira, for all her support, inspiration and example of perseverance, especially my aunt Ézia, my cousins Eleandro, Rafaela, Andressa and little Mariana, and the rest of the family.

Special thanks go to my advisor Dr Luciano Fleischfresser, who, as well as being a great friend, helped me a lot by sharing his knowledge so that I could carry out this work. To the other lecturers in the Environmental Engineering department. To the laboratory technician Mayko Cristian Sedoski for his great help in building the houses. To the other staff, especially the ladies responsible for cleaning the campus, for their patience when we got the corridors dirty a few times after they had cleaned them.

To the punks from Curitiba for all their motivation, to the guys from Serra Dourada. A huge thank you to the Methodist Church in Campo Mourão and Calvary Campo, for all the support, all the learning I've had and am having with you. To all my friends during graduation, the list would be huge here, but you know who you are, I will always carry the good memories of the moments with you. To all the people in the Comunic group, OQC, the Environmental Engineering Academic Centre, the English conversation group, DCE. To Rosana and Noé Esteves for always keeping the doors of the Terraço Grill open so that I could work there.

Last but not least, to the Lord Jesus Christ for finding me and giving my life real meaning and significance.

"And hope does not disappoint us, because God has poured his love into our hearts through the Holy Spirit he has given us."

(Romans 5: 5)

CONTENTS

CHAPTER 1

INTRODUCTION

Much has been said about sustainability, seeking to promote a balance between the use of available natural resources without jeopardising their existence, promoting social and economic benefits as well. The green roof is a sustainable alternative that has been implemented with the aim of reversing some of the negative environmental impacts of urbanisation.

According to Morais (2004), factors such as population density, intense land occupation, concentration of buildings, paving and the accumulation of pollutants in large urban centres have contributed significantly to the reduction of green spaces between built-up areas, generating various adverse environmental impacts. In addition, the reduction of these green areas can generate significant changes in thermal behaviour at the micro-scale.

Poorly planned urbanisation can cause impacts of great magnitude, severity and intensity. Considering this problem, it is necessary to develop techniques and technologies to minimise these impacts, seeking a balance between socio-economic activities and preserving the quality of the environment.

This accumulation of impermeable areas, caused by the accelerated process of urbanisation and development, has damaged the quality of life of people and the environment (UNHCR, 2008).

With the increase in built-up and paved areas there is also a reduction in the soil's water infiltration capacity. When this reduction is high, it leads to faster saturation of the soil, which in turn speeds up the surface runoff stage in the hydrological cycle. Costa (2012) and Neto (2012) emphasise the water retention capacity of green roofs and their contribution to cities with large impermeable areas that face river drainage problems.

Silva (2011) points out that green roofs have been adopted in many parts of the world, especially in Europe, as a means of minimising the problems caused by waterproofing in large cities. According to Neto (2012) there are many advantages to this technique, such as restoring the nature that was removed for the development of the city, visually improving the urban environment, improving the region's microclimate, increasing local biodiversity and creating a permeable surface that helps to reduce the load of rainwater that reaches the drainage networks.

According to Sailor (2012), green roofs also bring benefits such as reducing the magnitude of the urban heat island, promoting habitat, improving urban air quality, reducing noise transmission to buildings and reducing energy consumption in buildings. His work even quantified the savings made in electricity consumption through the use of green roofs.

Vacilikio (2011) emphasises that there are two ways of planting on roofs: intensive and extensive. The intensive type of roof requires deeper soil to support larger species and is usually installed on surfaces without slopes. Extensive roofs, on the other hand, require little soil as they can support creeping species and lawns.

CHAPTER 2

OBJECTIVES

2.1 GENERAL OBJECTIVE

The aim of this study was to assess the variation in internal temperatures in a wooden prototype with a grass-covered roof forming a green roof compared to a conventional fibre cement roof.

2.2 SPECIFIC OBJECTIVES

The specific objectives were:

- Construction of two prototype wooden rooms;
- Construction of three green roof modules with emerald grass, earth and expanded clay for one of the prototypes;
- Collecting data on the internal temperature of the tile and the temperature at the centre of both prototypes with 2 digital thermometers;
- Generate time series of the temperatures measured during the period of the experiments, from April 2014 to 31 December 2014.

July 2014, in order to compare the thermal behaviour of the roofs;

- Analysing the adaptation of the chosen grass to the green roof.

CHAPTER 3

LITERATURE REVIEW

Given the principles covered by Article 225 of the Brazilian constitution (1988) in relation to the environment, the responsibility of the public authorities and the community to defend and preserve it, and considering the importance of having an ecologically balanced environment, technological development has turned to alternatives that can reverse negative environmental impacts or minimise impacts from urbanisation. The green roof is an alternative for buildings that, according to Silva (2011), provides thermal and acoustic comfort in internal environments and has the function of increasing green areas, improving the environment and reducing the heat island effect in urban areas.

The green roof is an ancient construction technique that is still used today. According to Velazquez (2005), the combination of plants and architecture is not something new, as there are records of constructions with plants dating back to 2100 BC by the Ziggurats of ancient Mesopotamia, of Viking-era huts in Scotland and Ireland, and also the Hanging Gardens of Babylon (Figure 1), representing the oldest interpretation of a green roof, having been built between the 8th and 10th centuries BC.

The hanging gardens of Babylon were located on the east side of the Euphrates, in an ancient neighbourhood of the city, between the banks of the river and the royal palaces. It is one of the seven wonders of the world and the least

known since no trace of the monument has been found in the archaeological sites, with the only "suspicion" being a well outside the 15 standards that is thought to have been used to pump water (DISCOVERY CHANNEL, 2009 apud COSTA 2011).

Figure 1: Gardens of Babylon. Source: Socialphy (2012)

Looking at cultural and geographical issues, it is possible to see that the reason for planting vegetation on buildings was not always to obtain thermal advantages (the function of a green roof, for example), but often simply for aesthetic reasons.

Vecchia (2005), in his study comparing light green roofs with typical Brazilian roofs, analysed the thermal performance of green roofs in relation to other roofs with tiles such as ceramic, galvanised steel, 6 mm corrugated fibre cement and concrete slabs.

Vacilikio (2011) carried out a study comparing the variation in internal temperature in environments with green roofs and conventional roofs. These studies will serve as a reference for the research proposed in this study.

In a study on the application of the concept of sustainability, Schmidt (2009) pointed out various advantages of ecotiles over conventional roof tiles, such as improved air quality, protection for the building, rainfall advantages (water retention), contribution to biodiversity, acoustics (noise insulation), thermal comfort, food production, among others.

The subject is very vast and there is no doubt about the benefits the system brings to the environment and society, but measures must be adopted for the green roof to work properly, such as an adequate structure, remembering the need for structural calculations, because although the system is considered light there is still an overload on the structure, the waterproofing of the slab and an adequate drainage system

(SILVA, 2011).

3.1 TEMPERATURE CHANGES IN OUTDOOR ENVIRONMENTS AS A RESULT OF URBANISATION AND THERMAL DISCOMFORT

There are many problems facing society today as a result of poorly planned urbanisation. They can be visible, such as occupation in hazardous areas, or even risks that aren't always visible, such as air pollution.

Silva and Ribeiro (2006) carried out a study to analyse the temperature variations between a favela in the city of São Paulo and the Morumbi neighbourhood (the city's upmarket district). With 4 control points installed in the favela and 1 in a street in the Morumbi neighbourhood, the authors noted that the favela showed temperature extremes, reaching up to 3°C higher than the tree-lined street during the day, and 1°C lower at night.

Fatigati (2009) carried out a study of surface temperature variations in the city of São Paulo using satellite images from 1991 to 2006. According to the data and images obtained, the author pointed out that the most deprived areas of the city have the highest temperatures, thus creating microclimates of less thermal comfort for their inhabitants. Finally, the results indicate that this phenomenon of surface temperature variation is directly linked to anthropogenic action in the area.

In Rio Grande do Sul, Cllischonn (2005) used remote sensing and geoprocessing techniques to assess the different thermal environments in the Vale do Rio Pardo region. The author pointed out that urban areas have shown themselves to be warmer spots in the regional landscape as a whole, and that some areas are even clustering together to form a warmer plateau called a heat island.

Thermal discomfort often leads to higher electricity consumption by residents in affected areas. The Ministry of Cities (2005) therefore calls for the use of green roofs (figure 2) to minimise the temperature inside the house.

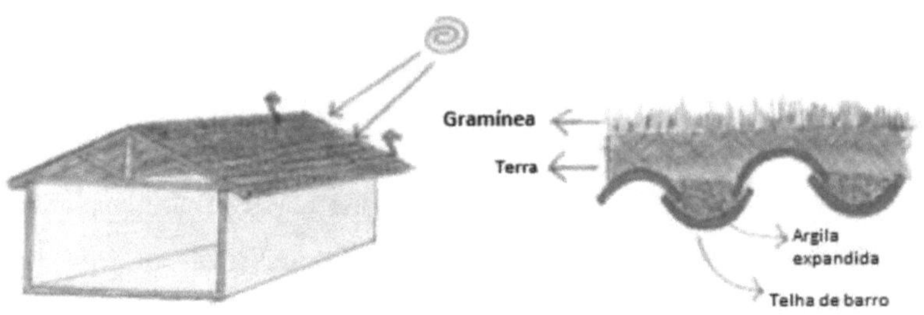

Figure 2: Example of a green roof. SOURCE: MINISTRY OF OF CITIES (2005)

3.2 USE OF GREEN ROOFS IN BUILDINGS.

The use of green roofs on buildings is not the most widely used type of roof, but it is gaining interest and space as more studies are carried out, pointing out the benefits of this technique. The implementation of a green roof on buildings depends directly on the type of slab present. There are two main uses for green roofs, one being to have a beautiful roof garden and the other to create an undisturbed habitat for flora and fauna in an urban area (INTERNATIONAL GREEN ROOF ASSOCIATION, 2013). Also taking into account bearable weight, maintenance, substrate, type of plant and costs, there are 3 types of green roof as shown in table 1: extensive, semi-extensive and intensive.

Table 1: Criteria for characterising green roof types

	Extensive	Semi-extensive	Intensive
Maintenance	Not much	Periodically	Quite a lot
Irrigation	No	Periodically	Regularly
Types of Vegetation	Mosses, sedum, herbs and grasses.	Grass, herbs and shrubs.	Lawns or perennials, shrubs e trees.
Height of System	60-200 mm	120-250 mm	150-400 mm
Weight	60-150 kg/m²	120 - 200 kg/m²	180-500 kg/m²
Cost	Bass	Medium	High
Utilisation	Ecological	Green Roof	Park like a garden

	protection layer	Designed	

Source: International Green Roof Association (2013)

CHAPTER 4

MATERIALS AND METHODS

4.1 CONSTRUCTION OF THE WOODEN ROOMS FOR THE EXPERIMENT

The work was carried out at the Federal Technological University of Paraná (UTFPR), Campo Mourão Campus. Firstly, 5cm x 10cm pine wood beams were used to build the structures for the two boxes. The structures were built in such a way as to be suitable for a sloping roof, as shown in figure 3a. Once the wooden structures were built, stain was applied, which is an impregnant to protect and embellish the wood, similar to varnish, in order to preserve it for the duration of the experiment, protecting it from humidity and fungi, as shown in figure 3b. Next, sheets of naval plywood were nailed to the sides and bottom of the floor of the prototypes. Finally, a door was made to allow access to the inside of the prototypes. Afterwards, another coat of stain was applied to help preserve the wood. Both prototypes were built with the same materials and then placed under the same conditions, 2 metres and 30 centimetres apart.

(a) (b)

Figure 3 - Structure of the prototype rooms used in the experiment: a) Construction phase, b) Application of stain **to the structures.**

As a result, both prototypes were 170 cm long, 170 cm wide, 170 cm high at the front and 210 cm high at the back.

To fix the thermometers inside the rooms, a support was designed to allow the thermocouples to reach the ceiling and the geometric centre of each box. These rooms were covered with fibre cement tiles fixed with nails. The green roof modules were built with dimensions of 244 cm long, 56 cm wide and 10 cm high. The three modules for the green roof were waterproofed with an asphalt blanket, then two holes were drilled and a small pipe with a radius of 2 cm was inserted into each hole to prevent water from accumulating at the bottom of the modules, as it is a sloping roof.

4.2 SETTING UP THE EXPERIMENT

The experiment was set up at the Campo Mourão Campus of the Federal Technological University of Paraná (UTFPR) . The plant species chosen for the green cover was Zoysia japonica from the Poaceae family, also known as Emerald grass.

Slabs of Esmeralda grass were purchased and placed on a layer of soil of approximately 6 cm so that the plant's roots could obtain nutrients and maintain their survival when placed on the green roof modules. A layer of expanded clay was also placed approximately 3 cm below the soil, as shown in Figure 4. Remember that one of the prototypes had the green roof modules on the roof, while the other only had fibre cement tiles. A little grated Styrofoam was also added to the earth to lighten the weight of the modules.

Figure 4: Expanded clay placed on the asphalt blanket to assemble the green roof modules.

Once the slabs of grass were on the green roof modules above the room, the grass was raked once a week if it didn't rain during this period, in order to help the grass adapt to the roof, since the earth substrate is smaller compared to the plant's original soil. Figure 5 shows the green roof modules still on the ground, as the Esmeralda grass slabs went through an observation period once they were placed on the modules, and Figure 5 also shows the grass slabs already adapted to the green roof modules, and superimposed on one of the rooms.

(a) (b)

Figure 5 - a): Emerald grass on one of the green roof modules, b) the grass slabs already adapted and on one of the rooms.

4.3 OBTAINING AND ANALYSING DATA

Two MT-600 digital thermometers with a capacity to store 10,000 data were used to obtain the data. Each one has 2 thermocouples, one of which is positioned to measure temperature variations in the fibre cement roof tile and is fixed to the inside of the roof tile, and the other is suspended in the geometric centre of the box to measure temperature variations in the internal air. The data obtained was downloaded using the thermometers' own software. External temperature and solar radiation data were supplied by the Automatic Meteorological Station at the UTFPR Campo Mourão campus. Data was collected every 30 minutes. A thermal paste was used to fix the thermocouples to the roof tiles, thus keeping them fixed (Figure 6).

Figura 6 - Thermometer positioned inside one of the prototypes, measuring the temperature of the tile and the temperature of the internal air.

This data was collected between April 2014 and May 2014. During this period, the variables measured were: internal air and roof temperatures of both prototypes, as shown in the diagram in figure 7. Solar radiation and external temperature data were provided by the automatic weather station at the UTFPR Campo Mourão campus.

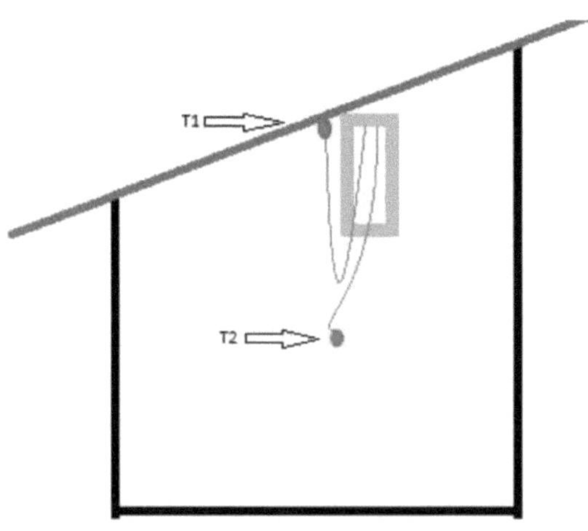

Figura 7 - Temperature collection diagram where T1 is the roof temperature and T2 the internal air temperature

From the data collected, graphs were generated using the Python 2.7 programming language and programme in order to demonstrate and compare the temperature evolution. The behaviour of the internal temperature variation in each prototype was then analysed, since both were subject to the same conditions, in order to identify the influence of the green roof on the internal temperature of one prototype compared to the other with a conventional roof. An example of this code for generating time series can be found in Appendix C.

CHAPTER 5

RESULTS AND DISCUSSION

5.1 COMPARISON OF INTERNAL TEMPERATURES BEFORE OVERLAPPING ONE OF THE ROOMS WITH THE GREEN ROOF

Figure 8 shows the internal air temperatures of both prototypes over 7 days while both had the same type of roof, i.e. before overlaying one of them with a green roof. It can be seen that the temperature values are within the thermometers' precision range, a factor that ensures the quality of the measurements.

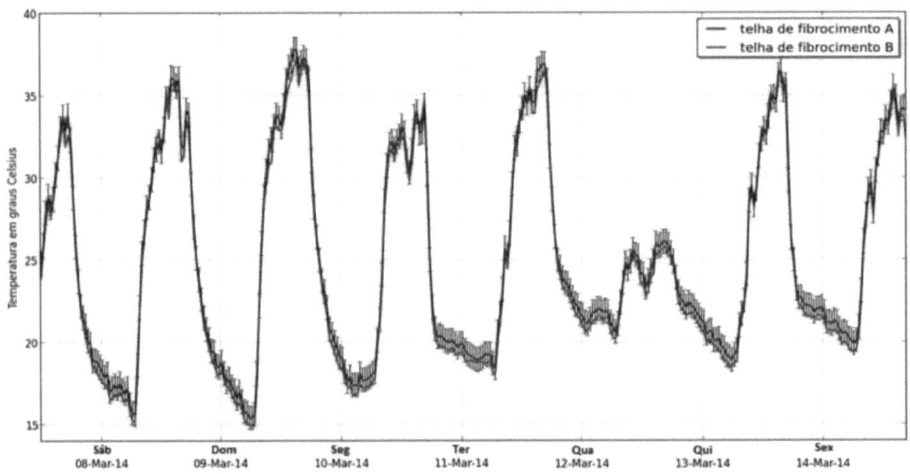

Figura 8 Measurements of thermocouples suspended inside the boxes. The accuracy of the thermometers is indicated in accordance with the technical specifications.

5.2 COMPARISON OF INTERNAL TEMPERATURES BETWEEN A ROOM WITH A GREEN ROOF AND ONE WITH FIBRE CEMENT ONLY

One week was chosen to represent the autumn season and another to represent winter when analysing the results. All the data used to generate the time series can be found in Appendices A and B.

Figure 9 shows the graph illustrating the temperature variations on 20 April

2014 for the asbestos-cement tile, the tile covered by the green roof and the outside temperature of the room. It can be seen that the conventional roof, only covered with the fibre cement tile, suffers greater temperature variations throughout the day, with both the highest temperature during the day and the lowest at night. The highest temperature reached by the roof that day was 34.1 °C, and the lowest was 13.5 °C, resulting in an amplitude of 20.6 °C. The roof overlaid by the green roof had a smaller range of variation, 16.7 °C. The highest temperature on the green roof that day was 31.7 °C, while the lowest was 15 °C. The range of variation in the outside temperature was 12.7 °C, with the lowest temperature being 14.5 °C and the highest 27.2 °C.

The internal air temperatures collected by the thermocouples suspended in the geometric centre of the prototypes also underwent similar variations, with the prototype with the conventional roof also showing a greater amplitude than the prototype with the green roof.

During the day, the prototype without the green roof had higher temperatures of up to 33.4 °C, and also cooled down more quickly as night fell to 14.6 °C, resulting in a temperature range of 18.8 °C. The prototype room with the green roof had a higher temperature of 32.2 °C and a lower temperature of 14.8 °C, resulting in a range of 17.4 °C (Figure 10).

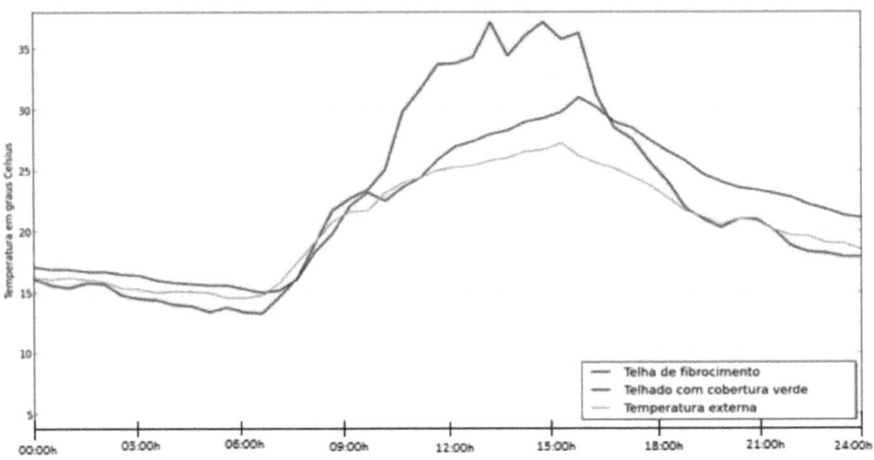

Figura 9- - Daily evolution of internal temperatures in fibre cement tiles without a green roof, with a green roof and the external temperature of the room.

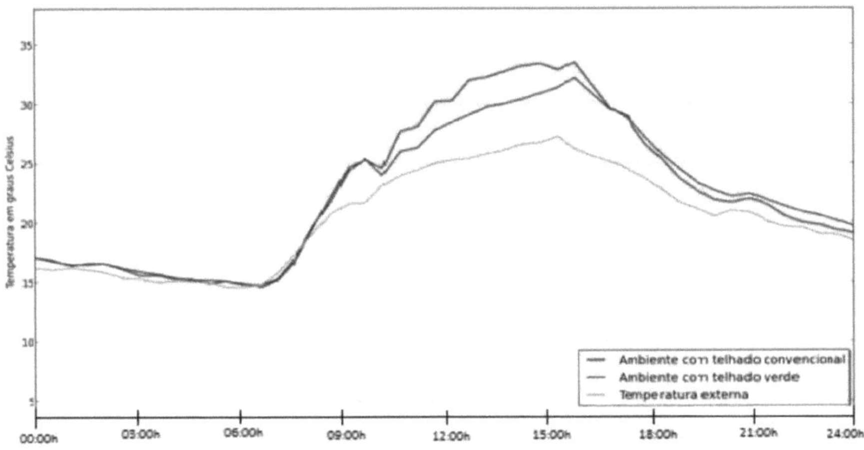

Figura 10- Daily evolution of the internal temperatures of the prototypes in relation to the external temperature.

Looking at the graphs containing the internal temperatures plus the external temperature, the results obtained are in line with other similar works such as Vecchia (2005) and Vacilikio (2011) who also noted that the green roof significantly reduced the maximum temperatures during the day, as well as taking longer to cool down at night. It's important to note that the walls of the prototypes were made of naval plywood, which is relatively thinner than the walls of ordinary masonry or even wooden houses. From the data obtained, it was possible to see that the temperatures under the roof tiles of both prototypes varied the most, i.e. they fluctuated over a wider range, mainly during the day when solar radiation is highest. Figure 11 shows the temperature variations of the roof tiles in the two rooms during a week in autumn, between 8 and 14 May 2014. It can be seen that the tile covered with the green roof has less variation in temperature. This is noticeable both during the day and at night; it remains in a smaller range compared to the fibre cement roof alone. The range of temperature variation of the green roof tile is between 13.6 and 32.6 °C, while the temperature of the fibre cement roof varies between 9.9 and 42 °C.

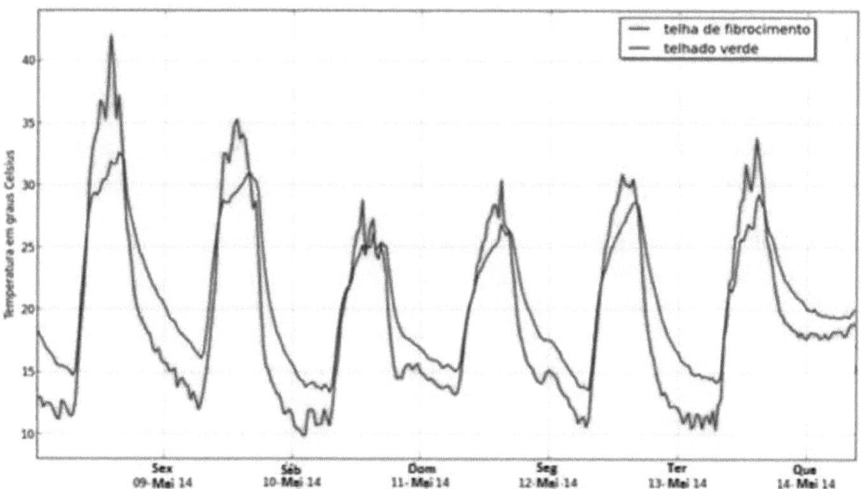

Figura 11- Temperature variation in the tiles of the room with the conventional roof and the one with the green roof from 8 to 14 May 2014.

In order to observe the behaviour of roof tile temperatures during the winter season, the week of 10 to 16 July was chosen, as shown in the figure 12.

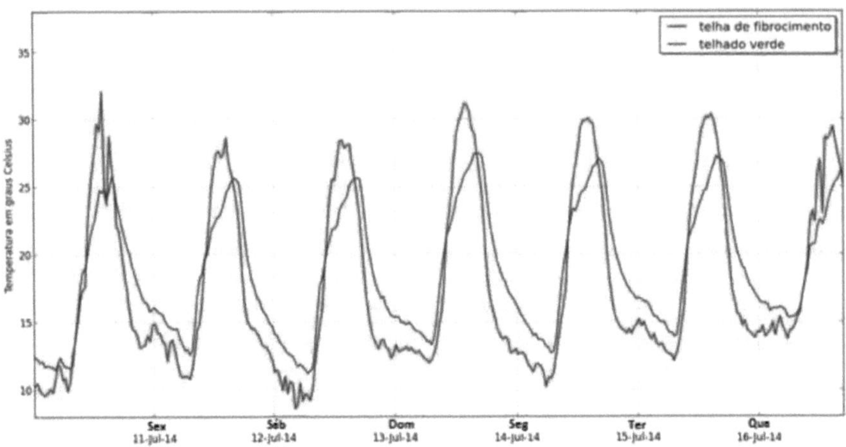

Figura 12- Variation in the internal temperatures of the roof tiles during the week of 10 to 16 July 2014.

The internal air temperatures measured with the thermocouples suspended in the centre of the rooms also showed differences throughout the experiment. Figure 13 shows the variation of the two internal temperatures in one week of the autumn season, from 8 to 14 May. Figure 14 shows the variation in temperatures from 10 to 16 July, covering part of the third and fourth week of the southern hemisphere's winter season. It can be seen that the most visible difference in variation is during the day when there is more solar radiation, between 10am and 3pm, and at night when there is no solar radiation over the experiment, generally between 9pm and 6am. It can be seen that the difference between the temperatures under the roof tiles is relatively greater than the difference between the internal air temperatures.

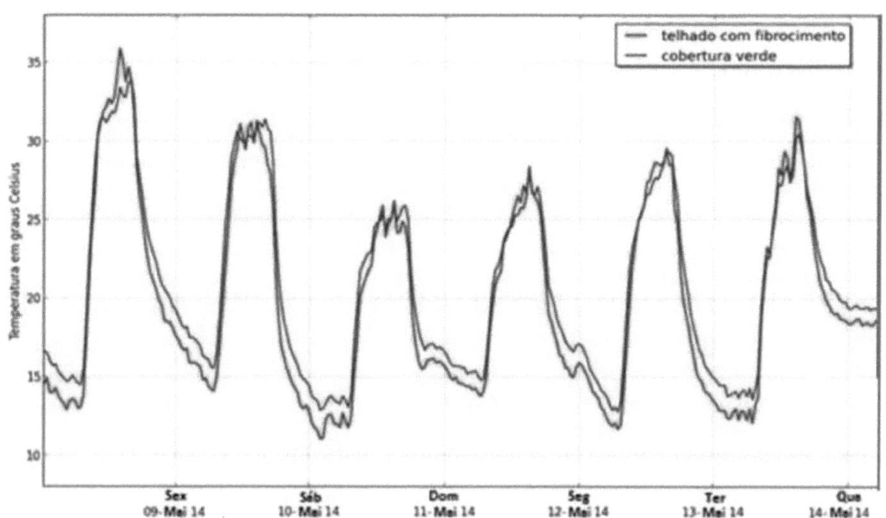

Figura 13 - Graph comparing the internal temperatures of the 2 rooms between 8 and 14 May 2014.

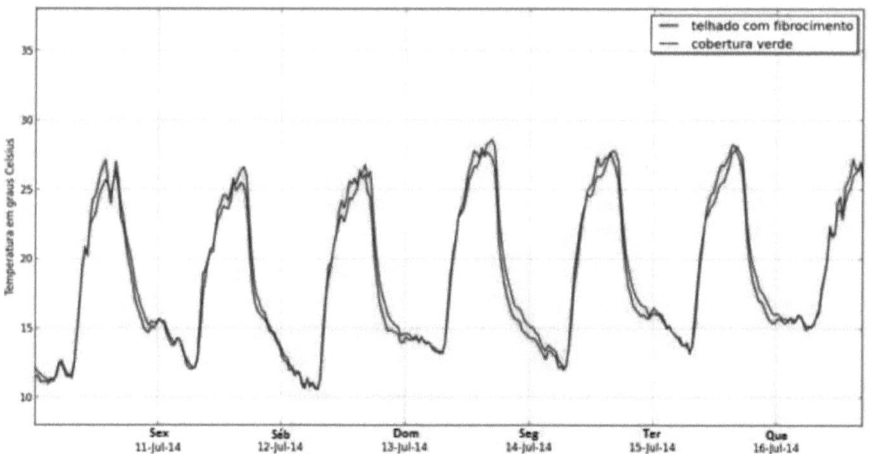

Figura 14 - Indoor temperatures measured between 10 and 16 July 2014.

It is worth emphasising once again that the walls used for the experiment are thin compared to a conventional house. Figure 15 shows how solar radiation directly hits both the roof and the walls of both prototypes. Rigão (2012) mentions that thermal inertia is characterised by the ability of a material to store and dissipate heat, and the speed at which a wall heats up or cools down is described as thermal inertia. The thickness of a wall is one of the main factors influencing its thermal inertia.

The small difference in indoor air temperature, especially during the winter week, can be explained by the low resistance of the walls in both rooms to the outside temperature, as well as the fact that solar radiation during this period was lower than at the start of the collections in April.

5.3 DIFFERENCES IN INTERNAL TEMPERATURES AND SOLAR RADIATION

Using the data obtained from the Automatic Weather Station on the campus itself, positioned a few metres from where the rooms for this experiment were placed (Figure 15), it was possible to generate time series comparing the difference in internal temperatures of each of the prototypes, i.e. temperature under the roof tile subtracted by the internal air temperature (T1 -T2), versus solar radiation. It was then possible to observe the behaviour of this difference over the course of the days according to variations in solar radiation.

Figura 15 - The experiment with one room covered with a green roof and the other with fibre cement tiles. In the background is the Campus Automatic Weather Station.

Figure 16 shows that T1-T2 (temperature under the tile minus the temperature of the internal air) of the prototype with the conventional roof tends to follow the flow of solar radiation. The temperature under the roof tile was higher than that of the internal air during periods of higher solar radiation. Consequently, the heat flow during these periods was from the roof tile to the geometric centre of the room (T1 to T2), so that the system entered thermal equilibrium.

As for the prototype covered with the green roof, the differences between T1 and T2 showed the opposite behaviour when subjected to the same solar radiation intensities. In other words, during periods of higher radiation during the day, the temperature of the tile with the green roof tended to be lower than the internal temperature of the same room.

This difference, contrary to the room with a conventional roof, indicates that the heat flow during these periods tended to be from the internal air in the geometric centre to the roof tile, from bottom to top (T2 to T1). This means that the difference between T1 and T2 was lower during periods of higher solar radiation, and the difference between T1 and T2 was higher during periods of zero radiation. In this way, the results show that at night the temperatures of the tiles in the room with the green roof were higher than in the other room, thus demonstrating the collaboration of the green roof in terms of heat retention. The heat flow during these periods of high solar radiation is illustrated in figure 17.

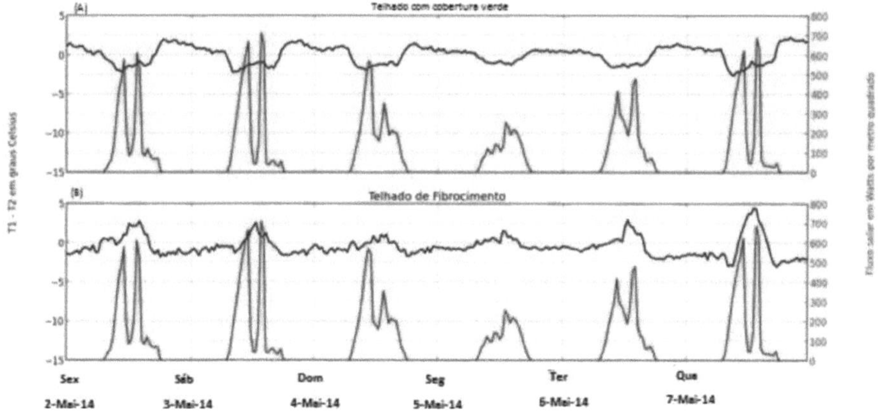

Figura 16 Difference between the temperature under the prototype roof and the

internal air temperature, in relation to the variation in solar radiation from 2 to 7 May. A) Prototype with green roof. B) Prototype without the green roof.

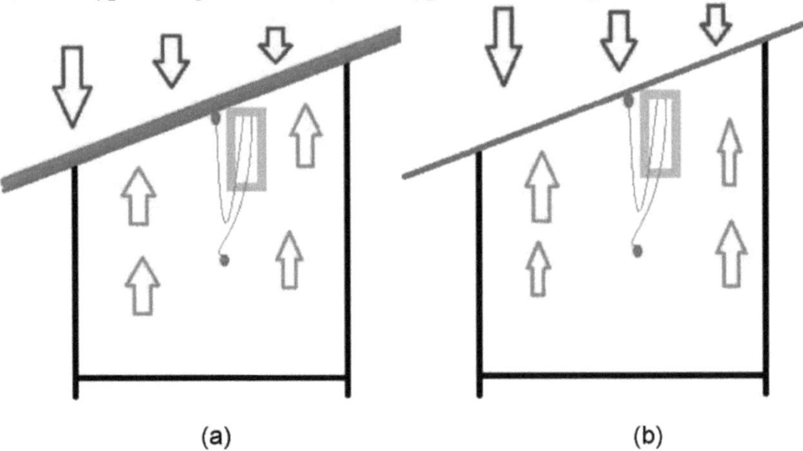

Figura 17 - Heat flow inside the prototypes during the periods of highest solar radiation: a) Prototype with the green roof, b) prototype without the green roof.

During periods of very low or zero solar radiation, there was also an opposite behaviour between the T_1 and T_2 differences of one prototype compared to the other. At night and in the early hours of the morning, the prototype clad only in fibre cement roof tiles showed a negative difference, which implies that the internal temperature was higher than the roof tile temperature. While the difference between T_1 and T_2 in the room covered with the green roof tended to show positive values in these periods of little or no solar radiation, which means that the roof tile temperature remained higher than the internal air temperature for most of this period.

5.4 ADAPTING THE GRASS TO THE ROOF SYSTEM

Even though the soil for the plant to grow in was shallow compared to its natural habitat, the grass chosen, known as Esmeralda, with the scientific name Zoysia japonica from the Poaceae family, did not show any resistance when it came to adapting to the green roof modules. As a precaution, the green roof was watered a few times when there were no or more days without rain. Although the roots had developed, the grass didn't need to be pruned during the experiment, which also makes it a good choice, as no maintenance was required.

CHAPTER 6

CONCLUSIONS

In carrying out this work, it was noted that the results obtained were favourable to the green roof, thus confirming some of the advantages mentioned in this work, as long as the system is planned and implemented correctly.

The data obtained showed that the green roof plays a role in attenuating the rise in internal temperatures during the day, when compared to the prototype with the fibre cement roof alone. This difference became clear on 24 April 2014, when the temperature of the prototype with the green roof was as much as 3.7 °C lower than the environment without the green roof, which had an internal temperature of 30.3 degrees Celsius.

During the experiment, it was observed that the room covered with the green roof took longer to cool down, as well as cooling down less. In the early hours of 10 May 2014, the green roof prototype had a temperature 2.2 °C higher than the other prototype, thus proving its efficiency. It was also noted that in the winter weeks, the differences in temperature from one prototype to the other were quite low.

During the four months that the experiment was carried out, the grass used proved to be suitable for this type of use, and once the system was in place it required little maintenance.

This work shows the advantages of using a green roof in relation to temperature variations.

REFERENCES

BRAZIL, 1988. **Constitution of the Federative Republic of Brazil. Article 225.** Available at at: <http://www.planalto.gov.br/ccivil_03/Constituicao/Constituicao.htm> Accessed on: 03/07/2013.

CLLISCHONN, Érica. **Surface Temperature and its Relation to Urbanisation in the Rio Pardo-RS Valley**. XI National Meeting of Graduate Studies and Research in Urban and Regional Planning - ANPUR. Salvador 23-27 May 2005. Bahia. Brazil. p. 2-12.

COSTA, Jefferson. COSTA, Anderson. POLETO, Cristiano. **Green Roof: Reducing and Delaying Surface Runoff**. REA - Revista de Estudos Ambientais. 2012. v. 14. n. 2esp. P. 50-56.

DISCOVERY, Channel. **The Seven Wonders of Antiquity**. 2009. Available at: <http://videos.neerkoli.com/video/VawFVfcBi8I/Seven-Wonders-of-the- Ancient-World-Discovery-Channel-Documentary.html> Accessed July 2013

FATIGATI, Flavio Laurenza. **Study of Surface Temperature Variation in the Municipality of São Paulo in the Period 1991-2006, Using Thermal Images from the LANDSAT-5 TM Satellite.** XIV.
Brazilian Symposium on Remote Sensing, Natal, Brazil, 25-29 April 2009. INPE, pg. 665- 662.

IGRA. International Green Roof Association. **Green Roof Types**. Available at:< http://igra-world.com/types_of_green_roofs/index.php> Accessed: July 2013.

MINISTRY OF CITIES. **Energy Efficiency in Social Interest Housing**. MCidades booklet. 2005

MORAIS, Carolina Santana de. **Thermal Performance of Vegetable Roofs in Buildings**

in the City of São Carlos/SP. 108 f. Master's dissertation. Department of Civil Engineering, Federal University of São Carlos, São Paulo, 2004.

NETO, Pedro de Souza Garrido. **Green Roofs Associated with Rainwater Harnessing Systems: Design of Two Prototypes for Future Studies on this Compensatory Technique in Urban Drainage and Sustainable Practice in Civil Construction.** Monograph. Federal University of Rio de Janeiro. Rio de Janeiro, 2012. p. 1, 44-90.

OHNUMA, Alfredo Akira Junior. **Unconventional Measures for Water Reservation and Water Pollution Control on Plots of Land**
Household. Thesis (Doctorate). São Carlos School of Engineering, University of São Paulo, 2008. p. 27-40

Rigão, Alessandro Onofre. **Behaviour of Small Structural Masonry Walls in the Face of High Temperatures**. Master's dissertation. Federal University of Santa Maria. Santa Maria- RS, 2012.

SAILOR, David J. ELLEY, Timothy B. GIBSON, Max. **Exploring the Building Energy Impacts of Green Roof Design Decisions - A Modelling Study of Buildings in 4 Distinct Climates**. Department of Mechanical and Materials Engineering, Portland State University, Portland. Journal of Building Physics. v. 35. 2012.

SCHMIDT, Franciele Taise Manica. **Application of the Sustainability Concept in a Single Family Residential Building - Case Study**. Monograph. Unijuí - Universidade Regional do Noroeste do Estado do Rio Grande do Sul. Department of Technology. Civil Engineering Course. Ijuí, November 2009.

SOCIALPHY. **Word's Seven Wonders - Babylon's Hanging Gardens.** Available at: <http://www.socialphy.com/posts/off-topic/9774/World_s- seven-wonders---Babylon_s-Hanging-gardens.html> Accessed: July 2013

SILVA, Edelci Nunes da. RIBEIRO, Helena. **Temperature Changes in Favela Outdoor Environments and Thermal Discomfort.** Department of Environmental Health. Faculty of Public Health. University of São Paulo, São Paulo, SP, Brazil. 2006

SILVA, Neusiane da Costa. **Green Roof: Building System with Greater Efficiency and Lower Environmental Impact.** Monograph. Specialisation course in Civil Construction at the UFMG School of Engineering. 2011.

VACILIKIO, Douglas V. **Comparison of Internal Temperatures in Rooms with Green and Conventional Roofs.** 2011. 42 f. Monograph (Higher Technology Course in Environmental Management) - Environmental Coordination, Federal Technological University of Paraná. Campo Mourão, 2011.

VECCHIA, Francisco. **Light Green Roof: Experimental Test**. ENCAC - ELACAC. Maceió, Alagoas, Brazil. 5 to 7 October 2005. p. 2-10.

VELASQUEZ, Linda S. **Organic Greenroof Architecture: Sustainable Design for the New Millennium.** Wiley Periodicals, Inc Environmental Quality Management/Summer 2005. p. 2-18

APPENDIX A

Table 1: Data obtained between 7 and 14 March 2014 from both rooms before the green roof was installed. T1 is the temperature close to the roof tile, and T2 is the air temperature in the centre of the room.

ROOF WITH GREEN ROOF	ROOF WITH FIBROCIMENTO

Date	Time	T1 °C	T2 °C	T1 °C	T2 °C
07/03/2014	11:30:00	23,9	26,2	24,8	26,5
07/03/2014	12:00:00	25,6	28,1	25,9	28,3
07/03/2014	12:30:00	27,3	31,2	27,9	31,5
07/03/2014	13:00:00	28,4	31,8	28,9	32
07/03/2014	13:30:00	27,6	29,5	28,2	29,4
07/03/2014	14:00:00	28,6	30,2	28,7	30,4
07/03/2014	14:30:00	29,9	36,4	30,2	36,4
07/03/2014	15:00:00	31,3	39,1	31,2	39,1
07/03/2014	15:30:00	32,4	39,3	33	39,6
07/03/2014	16:00:00	33,3	38,9	33,7	38,8
07/03/2014	16:30:00	31,9	36,1	32,9	36
07/03/2014	17:00:00	32,6	35,3	33,8	35,5
07/03/2014	17:30:00	31,8	32,5	32,3	32,3
07/03/2014	18:00:00	28,4	25,9	28,8	26
07/03/2014	18:30:00	25,4	22,8	25,8	23
07/03/2014	19:00:00	23,3	20,8	23,7	21,3
07/03/2014	19:30:00	21,9	19,5	22,2	19,9
07/03/2014	20:00:00	21	18,9	21,5	19,4
07/03/2014	20:30:00	20,3	18,3	21	18,7
07/03/2014	21:00:00	19,6	17,6	20,1	18,1
07/03/2014	21:30:00	19,4	17,6	19,9	18,1
07/03/2014	22:00:00	18,4	16,6	18,9	16,8
07/03/2014	22:30:00	18,6	17,2	18,9	17,6
07/03/2014	23:00:00	18,1	16,7	18,8	17,2
07/03/2014	23:30:00	17,8	16,3	18,5	16,8
08/03/2014	00:00:00	17,6	16,4	18,2	16,8
08/03/2014	00:30:00	17,4	16,1	17,9	16,4
08/03/2014	01:00:00	17,3	15,8	18,1	16,4
08/03/2014	01:30:00	16,5	14,4	17	14,9
08/03/2014	02:00:00	16,6	15	17,2	15,5
08/03/2014	02:30:00	16,9	15,4	17,4	16
	03:00:00	16,7	15	17,2	15,5

Table 1: Data obtained between 7 and 14 March 2014 from both rooms before the green roof was installed. T1 is the temperature in front of the roof tile, and T2 is the air temperature in the centre of the room. (continued)

	ROOF WITH GREEN ROOF			ROOF WITH FIBROCIMENTO	
Date	Time	T1 °C	T2 °C	T1 °C	T2 °C
11/03/2014	13:30:00	35,6	46,5	35,4	47,1
11/03/2014	14:00:00	34	38,7	34,6	39,3
11/03/2014	14:30:00	33,9	35,9	34,6	36,6
11/03/2014	15:00:00	35,4	40,6	36,5	41,6
11/03/2014	15:30:00	35,7	37,6	36,6	37,9
11/03/2014	16:00:00	36,1	41,7	36,9	42,1
11/03/2014	16:30:00	36,5	40,3	36,9	40,7
11/03/2014	17:00:00	35,3	35,9	35,9	35,8
11/03/2014	17:30:00	31,1	30	31,3	30,1
11/03/2014	18:00:00	28,7	27,3	28,7	27,3
11/03/2014	18:30:00	26,4	24,6	26,4	24,7
11/03/2014	19:00:00	24,7	22,9	25	23,1

11/03/2014	19:30:00	24	22,7	24,5	23,1
11/03/2014	20:00:00	23,4	22,5	23,9	22,7
11/03/2014	20:30:00	23,1	22,2	23,6	22,3
11/03/2014	21:00:00	23,1	22,1	23,5	22,5
11/03/2014	21:30:00	22,6	21,6	23,1	21,9
11/03/2014	22:00:00	22,3	21,1	22,7	21,4
11/03/2014	22:30:00	21,8	20,4	22,3	20,9
11/03/2014	23:00:00	21,6	20,5	22,2	20,7
11/03/2014	23:30:00	21,4	20	21,7	20,5
12/03/2014	00:00:00	21,1	19,7	21,6	20,2
12/03/2014	00:30:00	20,6	19,4	21,2	19,7
12/03/2014	01:00:00	20,8	19,5	21,1	20
12/03/2014	01:30:00	20,9	20,7	21,4	21
12/03/2014	02:00:00	21,3	21,1	21,8	21,4
12/03/2014	02:30:00	21,3	20,5	21,8	20,9
12/03/2014	03:00:00	21,6	20,9	22	21,3
12/03/2014	03:30:00	21,6	20,9	22	21,2
12/03/2014	04:00:00	21,3	20,5	21,8	20,8
12/03/2014	04:30:00	21,5	20,9	21,9	21,3
12/03/2014	05:00:00	21,5	20,2	21,8	20,6

Table 1: Data obtained between 7 and 14 March 2014 from both rooms before the green roof was installed. T1 is the temperature in front of the roof tile, and T2 is the air temperature in the centre of the room. (continued)

ROOF WITH GREEN ROOF			ROOF WITH FIBROCIMENTO	
Date	Time	T1 °C	T2 °C	T1 °C
08/03/2014	19:30:00	23,5	21,4	23,8
08/03/2014	20:00:00	22,1	20	22,6
08/03/2014	20:30:00	21	18,7	21,5
08/03/2014	21:00:00	20,4	18,4	20,8
08/03/2014	21:30:00	19,7	17,9	20,2
08/03/2014	22:00:00	19,1	17,1	19,5
08/03/2014	22:30:00	18,7	17,3	19,4
08/03/2014	23:00:00	18,2	16,6	18,6
08/03/2014	23:30:00	18	16,4	18,5
09/03/2014	00:00:00	18,3	16,8	18,8
09/03/2014	00:30:00	17,7	16	18,1
09/03/2014	01:00:00	17,3	15,7	17,6
09/03/2014	01:30:00	17,3	16,2	17,8
09/03/2014	02:00:00	16,8	15,1	17,3
09/03/2014	02:30:00	16,5	14,6	17,1
09/03/2014	03:00:00	16,1	14,4	16,6
09/03/2014	03:30:00	16,3	15,1	17
09/03/2014	04:00:00	15,9	14,2	16,4
09/03/2014	04:30:00	15,3	13,5	15,8
09/03/2014	05:00:00	15,1	13,7	15,7
09/03/2014	05:30:00	14,8	13,5	15,4
09/03/2014	06:00:00	14,7	13,1	15,4
09/03/2014	06:30:00	15,1	14,1	15,6
09/03/2014	07:00:00	17,7	16,2	18,1

09/03/2014	07:30:00	22	19,8	22,2
09/03/2014	08:00:00	25,8	23,7	26
09/03/2014	08:30:00	28,4	28,4	28,8
09/03/2014	09:00:00	29,7	31,3	30,4
09/03/2014	09:30:00	30,8	34	31,6
09/03/2014	10:00:00	30,9	35,4	32,3
09/03/2014	10:30:00	32,9	37,5	33,1
09/03/2014	11:00:00	33,1	37,8	33,8
09/03/2014	11:30:00	32,9	35,9	33,6

Table 1: Data obtained between 7 and 14 March 2014 from both rooms before the green roof was installed. T1 is the temperature in front of the roof tile, and T2 is the air temperature in the centre of the room. (continued)

	ROOF WITH GREEN ROOF			ROOF WITH FIBROCIMENTO	
Date	Time	T1 °C	T2 °C	T1 °C	T2 °C
11/03/2014	13:30:00	35,6	46,5	35,4	47,1
11/03/2014	14:00:00	34	38,7	34,6	39,3
11/03/2014	14:30:00	33,9	35,9	34,6	36,6
11/03/2014	15:00:00	35,4	40,6	36,5	41,6
11/03/2014	15:30:00	35,7	37,6	36,6	37,9
11/03/2014	16:00:00	36,1	41,7	36,9	42,1
11/03/2014	16:30:00	36,5	40,3	36,9	40,7
11/03/2014	17:00:00	35,3	35,9	35,9	35,8
11/03/2014	17:30:00	31,1	30	31,3	30,1
11/03/2014	18:00:00	28,7	27,3	28,7	27,3
11/03/2014	18:30:00	26,4	24,6	26,4	24,7
11/03/2014	19:00:00	24,7	22,9	25	23,1
11/03/2014	19:30:00	24	22,7	24,5	23,1
11/03/2014	20:00:00	23,4	22,5	23,9	22,7
11/03/2014	20:30:00	23,1	22,2	23,6	22,3
11/03/2014	21:00:00	23,1	22,1	23,5	22,5
11/03/2014	21:30:00	22,6	21,6	23,1	21,9
11/03/2014	22:00:00	22,3	21,1	22,7	21,4
11/03/2014	22:30:00	21,8	20,4	22,3	20,9
11/03/2014	23:00:00	21,6	20,5	22,2	20,7
11/03/2014	23:30:00	21,4	20	21,7	20,5
12/03/2014	00:00:00	21,1	19,7	21,6	20,2
12/03/2014	00:30:00	20,6	19,4	21,2	19,7
12/03/2014	01:00:00	20,8	19,5	21,1	20
12/03/2014	01:30:00	20,9	20,7	21,4	21
12/03/2014	02:00:00	21,3	21,1	21,8	21,4
12/03/2014	02:30:00	21,3	20,5	21,8	20,9
12/03/2014	03:00:00	21,6	20,9	22	21,3
12/03/2014	03:30:00	21,6	20,9	22	21,2
12/03/2014	04:00:00	21,3	20,5	21,8	20,8
12/03/2014	04:30:00	21,5	20,9	21,9	21,3
12/03/2014	05:00:00	21,5	20,2	21,8	20,6

Table 1: Data obtained between 7 and 14 March 2014 from both rooms before the green roof was installed. T1 is the temperature in front of the roof tile, and T2 is the air temperature in the centre of the room. (continued)

	ROOF WITH GREEN ROOF			ROOF WITH FIBROCIMENTO	
Date	Time	T1 °C	T2 °C	T1 °C	T2 °C
11/03/2014	13:30:00	35,6	46,5	35,4	47,1
11/03/2014	14:00:00	34	38,7	34,6	39,3
11/03/2014	14:30:00	33,9	35,9	34,6	36,6
11/03/2014	15:00:00	35,4	40,6	36,5	41,6
11/03/2014	15:30:00	35,7	37,6	36,6	37,9
11/03/2014	16:00:00	36,1	41,7	36,9	42,1
11/03/2014	16:30:00	36,5	40,3	36,9	40,7
11/03/2014	17:00:00	35,3	35,9	35,9	35,8
11/03/2014	17:30:00	31,1	30	31,3	30,1
11/03/2014	18:00:00	28,7	27,3	28,7	27,3
11/03/2014	18:30:00	26,4	24,6	26,4	24,7
11/03/2014	19:00:00	24,7	22,9	25	23,1
11/03/2014	19:30:00	24	22,7	24,5	23,1
11/03/2014	20:00:00	23,4	22,5	23,9	22,7
11/03/2014	20:30:00	23,1	22,2	23,6	22,3
11/03/2014	21:00:00	23,1	22,1	23,5	22,5
11/03/2014	21:30:00	22,6	21,6	23,1	21,9
11/03/2014	22:00:00	22,3	21,1	22,7	21,4
11/03/2014	22:30:00	21,8	20,4	22,3	20,9
11/03/2014	23:00:00	21,6	20,5	22,2	20,7
11/03/2014	23:30:00	21,4	20	21,7	20,5
12/03/2014	00:00:00	21,1	19,7	21,6	20,2
12/03/2014	00:30:00	20,6	19,4	21,2	19,7
12/03/2014	01:00:00	20,8	19,5	21,1	20
12/03/2014	01:30:00	20,9	20,7	21,4	21
12/03/2014	02:00:00	21,3	21,1	21,8	21,4
12/03/2014	02:30:00	21,3	20,5	21,8	20,9
12/03/2014	03:00:00	21,6	20,9	22	21,3
12/03/2014	03:30:00	21,6	20,9	22	21,2
12/03/2014	04:00:00	21,3	20,5	21,8	20,8
12/03/2014	04:30:00	21,5	20,9	21,9	21,3
12/03/2014	05:00:00	21,5	20,2	21,8	20,6

Table 1: Data obtained between 7 and 14 March 2014 from both rooms before the green roof was installed. T1 is the temperature close to the roof tile, and T2 is the air temperature in the centre of the room.(continued)

	ROOF WITH GREEN ROOF			ROOF WITH FIBROCIMENTO	
Date	Time	T1 °C	T2 °C	T1 °C	T2 °C
11/03/2014	13:30:00	35,6	46,5	35,4	47,1
11/03/2014	14:00:00	34	38,7	34,6	39,3
11/03/2014	14:30:00	33,9	35,9	34,6	36,6
11/03/2014	15:00:00	35,4	40,6	36,5	41,6

11/03/2014	15:30:00	35,7	37,6	36,6	37,9
11/03/2014	16:00:00	36,1	41,7	36,9	42,1
11/03/2014	16:30:00	36,5	40,3	36,9	40,7
11/03/2014	17:00:00	35,3	35,9	35,9	35,8
11/03/2014	17:30:00	31,1	30	31,3	30,1
11/03/2014	18:00:00	28,7	27,3	28,7	27,3
11/03/2014	18:30:00	26,4	24,6	26,4	24,7
11/03/2014	19:00:00	24,7	22,9	25	23,1
11/03/2014	19:30:00	24	22,7	24,5	23,1
11/03/2014	20:00:00	23,4	22,5	23,9	22,7
11/03/2014	20:30:00	23,1	22,2	23,6	22,3
11/03/2014	21:00:00	23,1	22,1	23,5	22,5
11/03/2014	21:30:00	22,6	21,6	23,1	21,9
11/03/2014	22:00:00	22,3	21,1	22,7	21,4
11/03/2014	22:30:00	21,8	20,4	22,3	20,9
11/03/2014	23:00:00	21,6	20,5	22,2	20,7
11/03/2014	23:30:00	21,4	20	21,7	20,5
12/03/2014	00:00:00	21,1	19,7	21,6	20,2
12/03/2014	00:30:00	20,6	19,4	21,2	19,7
12/03/2014	01:00:00	20,8	19,5	21,1	20
12/03/2014	01:30:00	20,9	20,7	21,4	21
12/03/2014	02:00:00	21,3	21,1	21,8	21,4
12/03/2014	02:30:00	21,3	20,5	21,8	20,9
12/03/2014	03:00:00	21,6	20,9	22	21,3
12/03/2014	03:30:00	21,6	20,9	22	21,2
12/03/2014	04:00:00	21,3	20,5	21,8	20,8
12/03/2014	04:30:00	21,5	20,9	21,9	21,3
12/03/2014	05:00:00	21,5	20,2	21,8	20,6

Table 1: Data obtained between 7 and 14 March 2014 from both rooms before the green roof was installed. T1 is the temperature close to the roof tile, and T2 is the air temperature in the centre of the room.(continued)

	ROOF WITH GREEN ROOF			ROOF WITH FIBROCIMENTO	
Date	Time	T1 °C	T2 °C	T1 °C	T2 °C
11/03/2014	13:30:00	35,6	46,5	35,4	47,1
11/03/2014	14:00:00	34	38,7	34,6	39,3
11/03/2014	14:30:00	33,9	35,9	34,6	36,6
11/03/2014	15:00:00	35,4	40,6	36,5	41,6
11/03/2014	15:30:00	35,7	37,6	36,6	37,9
11/03/2014	16:00:00	36,1	41,7	36,9	42,1
11/03/2014	16:30:00	36,5	40,3	36,9	40,7
11/03/2014	17:00:00	35,3	35,9	35,9	35,8
11/03/2014	17:30:00	31,1	30	31,3	30,1
11/03/2014	18:00:00	28,7	27,3	28,7	27,3
11/03/2014	18:30:00	26,4	24,6	26,4	24,7
11/03/2014	19:00:00	24,7	22,9	25	23,1
11/03/2014	19:30:00	24	22,7	24,5	23,1
11/03/2014	20:00:00	23,4	22,5	23,9	22,7
11/03/2014	20:30:00	23,1	22,2	23,6	22,3

11/03/2014	21:00:00	23,1	22,1	23,5	22,5
11/03/2014	21:30:00	22,6	21,6	23,1	21,9
11/03/2014	22:00:00	22,3	21,1	22,7	21,4
11/03/2014	22:30:00	21,8	20,4	22,3	20,9
11/03/2014	23:00:00	21,6	20,5	22,2	20,7
11/03/2014	23:30:00	21,4	20	21,7	20,5
12/03/2014	00:00:00	21,1	19,7	21,6	20,2
12/03/2014	00:30:00	20,6	19,4	21,2	19,7
12/03/2014	01:00:00	20,8	19,5	21,1	20
12/03/2014	01:30:00	20,9	20,7	21,4	21
12/03/2014	02:00:00	21,3	21,1	21,8	21,4
12/03/2014	02:30:00	21,3	20,5	21,8	20,9
12/03/2014	03:00:00	21,6	20,9	22	21,3
12/03/2014	03:30:00	21,6	20,9	22	21,2
12/03/2014	04:00:00	21,3	20,5	21,8	20,8
12/03/2014	04:30:00	21,5	20,9	21,9	21,3
12/03/2014	05:00:00	21,5	20,2	21,8	20,6

Table 1: Data obtained between 7 and 14 March 2014 from both rooms before the green roof was installed. T1 is the temperature in front of the roof tile and T2 is the air temperature in the centre of the room.

	ROOF WITH GREEN ROOF			ROOF WITH FIBROCIMENTO	
Date	Time	T1 °C	T2 °C	T1 °C	T2 °C
11/03/2014	13:30:00	35,6	46,5	35,4	47,1
11/03/2014	14:00:00	34	38,7	34,6	39,3
11/03/2014	14:30:00	33,9	35,9	34,6	36,6
11/03/2014	15:00:00	35,4	40,6	36,5	41,6
11/03/2014	15:30:00	35,7	37,6	36,6	37,9
11/03/2014	16:00:00	36,1	41,7	36,9	42,1
11/03/2014	16:30:00	36,5	40,3	36,9	40,7
11/03/2014	17:00:00	35,3	35,9	35,9	35,8
11/03/2014	17:30:00	31,1	30	31,3	30,1
11/03/2014	18:00:00	28,7	27,3	28,7	27,3
11/03/2014	18:30:00	26,4	24,6	26,4	24,7
11/03/2014	19:00:00	24,7	22,9	25	23,1
11/03/2014	19:30:00	24	22,7	24,5	23,1
11/03/2014	20:00:00	23,4	22,5	23,9	22,7
11/03/2014	20:30:00	23,1	22,2	23,6	22,3
11/03/2014	21:00:00	23,1	22,1	23,5	22,5
11/03/2014	21:30:00	22,6	21,6	23,1	21,9
11/03/2014	22:00:00	22,3	21,1	22,7	21,4
11/03/2014	22:30:00	21,8	20,4	22,3	20,9
11/03/2014	23:00:00	21,6	20,5	22,2	20,7
11/03/2014	23:30:00	21,4	20	21,7	20,5
12/03/2014	00:00:00	21,1	19,7	21,6	20,2
12/03/2014	00:30:00	20,6	19,4	21,2	19,7
12/03/2014	01:00:00	20,8	19,5	21,1	20
12/03/2014	01:30:00	20,9	20,7	21,4	21
12/03/2014	02:00:00	21,3	21,1	21,8	21,4

12/03/2014	02:30:00	21,3	20,5	21,8	20,9
12/03/2014	03:00:00	21,6	20,9	22	21,3
12/03/2014	03:30:00	21,6	20,9	22	21,2
12/03/2014	04:00:00	21,3	20,5	21,8	20,8
12/03/2014	04:30:00	21,5	20,9	21,9	21,3
12/03/2014	05:00:00	21,5	20,2	21,8	20,6

Table 1: Data obtained between 7 and 14 March 2014 from both rooms before the green roof was installed. T1 is the temperature close to the roof tile, and T2 is the air temperature in the centre of the room.(continued)

	ROOF WITH GREEN ROOF			ROOF WITH FIBROCIMENTO	
Date	Time	T1 °C	T2 °C	T1 °C	T2 °C
11/03/2014	13:30:00	35,6	46,5	35,4	47,1
11/03/2014	14:00:00	34	38,7	34,6	39,3
11/03/2014	14:30:00	33,9	35,9	34,6	36,6
11/03/2014	15:00:00	35,4	40,6	36,5	41,6
11/03/2014	15:30:00	35,7	37,6	36,6	37,9
11/03/2014	16:00:00	36,1	41,7	36,9	42,1
11/03/2014	16:30:00	36,5	40,3	36,9	40,7
11/03/2014	17:00:00	35,3	35,9	35,9	35,8
11/03/2014	17:30:00	31,1	30	31,3	30,1
11/03/2014	18:00:00	28,7	27,3	28,7	27,3
11/03/2014	18:30:00	26,4	24,6	26,4	24,7
11/03/2014	19:00:00	24,7	22,9	25	23,1
11/03/2014	19:30:00	24	22,7	24,5	23,1
11/03/2014	20:00:00	23,4	22,5	23,9	22,7
11/03/2014	20:30:00	23,1	22,2	23,6	22,3
11/03/2014	21:00:00	23,1	22,1	23,5	22,5
11/03/2014	21:30:00	22,6	21,6	23,1	21,9
11/03/2014	22:00:00	22,3	21,1	22,7	21,4
11/03/2014	22:30:00	21,8	20,4	22,3	20,9
11/03/2014	23:00:00	21,6	20,5	22,2	20,7
11/03/2014	23:30:00	21,4	20	21,7	20,5
12/03/2014	00:00:00	21,1	19,7	21,6	20,2
12/03/2014	00:30:00	20,6	19,4	21,2	19,7
12/03/2014	01:00:00	20,8	19,5	21,1	20
12/03/2014	01:30:00	20,9	20,7	21,4	21
12/03/2014	02:00:00	21,3	21,1	21,8	21,4
12/03/2014	02:30:00	21,3	20,5	21,8	20,9
12/03/2014	03:00:00	21,6	20,9	22	21,3
12/03/2014	03:30:00	21,6	20,9	22	21,2
12/03/2014	04:00:00	21,3	20,5	21,8	20,8
12/03/2014	04:30:00	21,5	20,9	21,9	21,3
12/03/2014	05:00:00	21,5	20,2	21,8	20,6

Table 1: Data obtained between 7 and 14 March 2014 from both rooms before the green roof was installed. T1 is the temperature in front of the roof tile and T2 is the air temperature in the centre of the room.

	ROOF WITH GREEN ROOF			ROOF WITH FIBROCIMENTO	
Date	Time	T1 °C	T2 °C	T1 °C	T2 °C
11/03/2014	13:30:00	35,6	46,5	35,4	47,1
11/03/2014	14:00:00	34	38,7	34,6	39,3
11/03/2014	14:30:00	33,9	35,9	34,6	36,6
11/03/2014	15:00:00	35,4	40,6	36,5	41,6
11/03/2014	15:30:00	35,7	37,6	36,6	37,9
11/03/2014	16:00:00	36,1	41,7	36,9	42,1
11/03/2014	16:30:00	36,5	40,3	36,9	40,7
11/03/2014	17:00:00	35,3	35,9	35,9	35,8
11/03/2014	17:30:00	31,1	30	31,3	30,1
11/03/2014	18:00:00	28,7	27,3	28,7	27,3
11/03/2014	18:30:00	26,4	24,6	26,4	24,7
11/03/2014	19:00:00	24,7	22,9	25	23,1
11/03/2014	19:30:00	24	22,7	24,5	23,1
11/03/2014	20:00:00	23,4	22,5	23,9	22,7
11/03/2014	20:30:00	23,1	22,2	23,6	22,3
11/03/2014	21:00:00	23,1	22,1	23,5	22,5
11/03/2014	21:30:00	22,6	21,6	23,1	21,9
11/03/2014	22:00:00	22,3	21,1	22,7	21,4
11/03/2014	22:30:00	21,8	20,4	22,3	20,9
11/03/2014	23:00:00	21,6	20,5	22,2	20,7
11/03/2014	23:30:00	21,4	20	21,7	20,5
12/03/2014	00:00:00	21,1	19,7	21,6	20,2
12/03/2014	00:30:00	20,6	19,4	21,2	19,7
12/03/2014	01:00:00	20,8	19,5	21,1	20
12/03/2014	01:30:00	20,9	20,7	21,4	21
12/03/2014	02:00:00	21,3	21,1	21,8	21,4
12/03/2014	02:30:00	21,3	20,5	21,8	20,9
12/03/2014	03:00:00	21,6	20,9	22	21,3
12/03/2014	03:30:00	21,6	20,9	22	21,2
12/03/2014	04:00:00	21,3	20,5	21,8	20,8
12/03/2014	04:30:00	21,5	20,9	21,9	21,3
12/03/2014	05:00:00	21,5	20,2	21,8	20,6

Table 1: Data obtained between 7 and 14 March 2014 from both rooms before the green roof was installed. T1 is the temperature in front of the roof tile, and T2 is the air temperature in the centre of the room (conclusion).

	ROOF WITH GREEN ROOF			ROOF WITH FIBROCIMENTO	
Date	Time	T1 °C	T2 °C	T1 °C	T2 °C
11/03/2014	13:30:00	35,6	46,5	35,4	47,1
11/03/2014	14:00:00	34	38,7	34,6	39,3
11/03/2014	14:30:00	33,9	35,9	34,6	36,6
11/03/2014	15:00:00	35,4	40,6	36,5	41,6
11/03/2014	15:30:00	35,7	37,6	36,6	37,9
11/03/2014	16:00:00	36,1	41,7	36,9	42,1
11/03/2014	16:30:00	36,5	40,3	36,9	40,7
11/03/2014	17:00:00	35,3	35,9	35,9	35,8

11/03/2014	17:30:00	31,1	30	31,3	30,1
11/03/2014	18:00:00	28,7	27,3	28,7	27,3
11/03/2014	18:30:00	26,4	24,6	26,4	24,7
11/03/2014	19:00:00	24,7	22,9	25	23,1
11/03/2014	19:30:00	24	22,7	24,5	23,1
11/03/2014	20:00:00	23,4	22,5	23,9	22,7
11/03/2014	20:30:00	23,1	22,2	23,6	22,3
11/03/2014	21:00:00	23,1	22,1	23,5	22,5
11/03/2014	21:30:00	22,6	21,6	23,1	21,9
11/03/2014	22:00:00	22,3	21,1	22,7	21,4
11/03/2014	22:30:00	21,8	20,4	22,3	20,9
11/03/2014	23:00:00	21,6	20,5	22,2	20,7

APPENDIX B

Table 2: Data obtained between 8 and 14 May 2014 from both rooms after the green roof was installed. T1 is the temperature close to the roof tile, T2 the air temperature in the centre of the room and Te the outside temperature. (continued)

	green roof		fibre cement		
Date	T1 °C	T2 °C	T1 °C	T2 °C	Te
08-May-14-00:08:50	12,9	14,6	18,2	16,6	15,07
08-May-14-00:38:50	13	14,9	17,8	16,5	14,71
08-May-14-01:08:50	12,2	14	17,4	16	15,17
08-May-14-01:38:50	12,5	14	16,9	15,7	14,45
08-May-14-02:08:50	12,5	14,3	16,7	15,8	14,72
08-May-14-02:38:50	12,4	13,8	16,4	15,3	14,75
08-May-14-03:08:50	11,9	13,5	16,1	15,1	14,13
08-May-14-03:38:50	11,4	13,3	15,6	14,9	13,79
08-May-14-04:08:50	11,2	12,9	15,5	14,7	13,5
08-May-14-04:38:50	12,7	13,5	15,5	14,8	13,14
08-May-14-05:08:50	12,5	13,6	15,4	15,1	14,16
08-May-14-05:38:50	12	13,5	15,3	14,8	14,38
08-May-14-06:08:50	11,5	13	15,1	14,6	14,14
08-May-14-06:38:50	11,5	13,1	14,7	14,5	13,56
08-May-14-07:08:50	12,5	13,9	15,1	15,6	13,52
08-May-14-07:38:50	15,5	17,5	17,5	18,9	14,17
08-May-14-08:08:50	18,9	21,4	20,7	22,8	17,15
08-May-14-08:38:50	21,4	24,5	23,4	25,8	19,83
08-May-14-09:08:50	24,9	26,9	26	28,6	21,56
08-May-14-09:38:50	30,2	30,3	28	30,6	23,43
08-May-14-10:08:50	32,6	31,2	29,1	31,4	25,74
08-May-14-10:38:50	33,8	31,9	29,4	31,5	26,31
08-May-14-11:08:50	34,5	32,2	29,2	31,2	26,4
08-May-14-11:38:50	36,8	32,7	29,6	31,5	26,37
08-May-14-12:08:50	36,4	32,4	30,4	31,9	26,73
08-May-14-12:38:50	35,3	32,8	30,4	31,8	27,43
08-May-14-13:08:50	37,9	34,4	30,9	32,4	27,01
08-May-14-13:38:50	42	35,9	31,9	33,4	27,75
08-May-14-14:08:50	39,7	35,1	31,7	32,9	28,4
08-May-14-14:38:50	35,4	33,8	31,7	32,8	27,97
08-May-14-15:08:50	37,2	34,7	32,6	33,7	27,89
08-May-14-15:38:50	34,4	33,9	32,4	33,9	27,92
08-May-14-16:08:50	30,4	31,9	32	32,8	27,84

Table 2: Data obtained between 8 and 14 May 2014 from both rooms after the green roof was installed. T1 is the temperature in front of the roof tile, T2 the air temperature in the centre of the room and Te the outside temperature.(continued)

	Green Roof		Fibre cement roof		
Date and time	T1 °C	T2 °C	T1 °C	T2 °C	Te
08-May-14-16:38:50	27	28,2	29,4	29,2	26,59
08-May-14-17:08:50	25,6	26,6	28,1	27,8	25,08
08-May-14-17:38:50	23,5	25,3	27,2	26,6	24,51
08-May-14-18:08:50	21,3	23,7	26,3	25,2	23,98
08-May-14-18:38:50	20,2	22,5	25,4	24,1	23,03
08-May-14-19:08:50	19,5	21,6	24,6	23,4	22,29
08-May-14-19:38:50	19,1	21,1	24,2	22,8	21,61
08-May-14-20:08:50	18,6	20,6	23,5	22,3	21,19
08-May-14-20:38:50	18,2	19,9	23	21,5	20,69
08-May-14-21:08:50	17,9	19,6	22,4	20,9	19,78
08-May-14-21:38:50	16,9	18,6	21,9	20,6	19,16
08-May-14-22:08:50	16,6	18,5	21,6	20,3	18,63
08-May-14-22:38:50	16,9	18,5	21	20	18,69
08-May-14-23:08:50	16	18,1	20,8	19,5	19,04
08-May-14-23:38:50	15,7	17,6	20,4	19,3	18,77
09-May-14-00:08:50	15,7	17,4	20	18,9	18,12
09-May-14-00:38:50	15,1	16,9	19,3	18,4	17,87
09-May-14-01:08:50	15,1	16,7	19,2	18,1	17,4
09-May-14-01:38:50	15,3	16,8	19	18,2	17,22
09-May-14-02:08:50	13,8	15,9	18,5	17,5	17
09-May-14-02:38:50	14,4	15,9	18,1	17,5	16,38
09-May-14-03:08:50	14,5	15,9	18	17,4	16,51
09-May-14-03:38:50	14	15,8	17,7	17,2	16,69
09-May-14-04:08:50	13,9	15,6	17,5	16,9	16,26
09-May-14-04:38:50	12,8	14,8	17,3	16,3	15,88
09-May-14-05:08:50	13,4	14,9	16,9	16,2	15,68
09-May-14-05:38:50	12,8	14,4	16,6	16,1	15,5
09-May-14-06:08:50	12	14,2	16,4	15,6	15,02
09-May-14-06:38:50	12,6	14,1	16,1	15,6	14,92
09-May-14-07:08:50	13,7	15,1	16,6	16,6	15,05
09-May-14-07:38:50	15,1	16,9	18,3	18,9	15,45
09-May-14-08:08:50	17	19,4	20,4	21,7	16,43

Table 2: Data obtained between 8 and 14 May 2014 from both rooms after the green roof was installed. T1 is the temperature close to the roof tile, T2 the air temperature in the centre of the room and Te the outside temperature.(continued)

	Green Roof		Fibre cement roof		
Date and time	T1 °C	T2 °C	T1 °C	T2 °C	Te
09-May-14-08:38:50	19,5	22	23	24,7	17,78
09-May-14-09:08:50	21,7	24,6	25	26,8	19,62
09-May-14-09:38:50	26	27,3	26,8	28,9	21,08
09-May-14-10:08:50	28,8	28,7	27,9	29,8	22,48
09-May-14-10:38:50	32,5	29,5	28,8	30,6	23,16
09-May-14-11:08:50	32,5	31,1	28,6	30,2	23,99
09-May-14-11:38:50	31,8	30,4	28,6	30	23,9
09-May-14-12:08:50	33,2	29,5	29,1	29,8	23,86
09-May-14-12:38:50	34,9	30,9	29,3	30,4	24,11
09-May-14-13:08:50	35,3	31,2	29,5	30,3	24,59
09-May-14-13:38:50	33,7	29,9	29,8	30	24,89
09-May-14-14:08:50	34,1	31,3	30,1	30,9	25,12
09-May-14-14:38:50	33,6	30,5	30,5	31,2	25,14
09-May-14-15:08:50	31,6	29,8	30,9	30,9	24,97

09-May-14-15:38:50	30,1	29,6	30,9	31,4	24,69
09-May-14-16:08:50	28,1	28,5	30,5	30,8	23,96
09-May-14-16:38:50	28,7	28	30,3	30,6	23,6
09-May-14-17:08:50	25	25,6	29,4	28,6	22,61
09-May-14-17:38:50	20,4	21,6	26,4	24,3	21,42
09-May-14-18:08:50	17,1	18,9	24,1	21,6	19,86
09-May-14-18:38:50	15,5	17	22,3	19,8	18,25
09-May-14-19:08:50	15,1	16,3	20,9	18,6	17,12
09-May-14-19:38:50	14,3	15,6	19,9	17,8	16,64
09-May-14-20:08:50	13,9	15,1	19,1	17	16,25
09-May-14-20:38:50	13,3	14,6	18,2	16,5	15,4
09-May-14-21:08:50	13,2	14,3	17,8	16,1	14,93
09-May-14-21:38:50	12,5	14	17,2	15,7	14,78
09-May-14-22:08:50	11,6	13,2	16,6	15,1	14,71
09-May-14-22:38:50	11,8	13	16,3	14,9	14,22
09-May-14-23:08:50	12	13,2	15,9	14,7	14,12
09-May-14-23:38:50	11,6	12,9	15,6	14,5	14,08
10-May-14-00:08:50	10,6	12,1	15,2	13,8	13,98
10-May-14-00:38:50	10,4	11,8	14,7	13,6	13,58

Table 2: Data obtained between 8 and 14 May 2014 from both rooms after the green roof was installed. T1 is the temperature in front of the roof tile, T2 the air temperature in the centre of the room and Te the outside temperature.(continued)

	Green Roof		Fibre cement roof		
Date and time	T1 °C	T2 °C	T1 °C	T2 °C	Te
10-May-14-01:08:50	10,3	11,6	14,4	13,5	13,16
10-May-14-01:38:50	9,9	11,1	14,3	12,9	12,74
10-May-14-02:08:50	9,9	11,1	13,8	13	12,42
10-May-14-02:38:50	11,9	12,3	14	13,3	12,36
10-May-14-03:08:50	12	12,6	14,1	13,7	13,26
10-May-14-03:38:50	11,9	12,6	14,2	13,8	13,69
10-May-14-04:08:50	10,7	12	13,8	13,5	13,53
10-May-14-04:38:50	10,8	12	13,8	13,4	13,25
10-May-14-05:08:50	10,8	11,8	13,6	13,3	13,52
10-May-14-05:38:50	12	12,7	14	13,8	13,6
10-May-14-06:08:50	11,3	12,2	13,8	13,5	13,69
10-May-14-06:38:50	10,7	11,8	13,4	13,1	12,7
10-May-14-07:08:50	12,1	12,5	14	14	12,22
10-May-14-07:38:50	14,5	15,2	16,1	17,2	12,69
10-May-14-08:08:50	16,1	17,3	18,2	19,5	14,15
10-May-14-08:38:50	18,9	20,1	20	21,7	15,46
10-May-14-09:08:50	20,1	20,8	20,9	22,2	16,68
10-May-14-09:38:50	21,3	21,3	21,6	22,6	17,19
10-May-14-10:08:50	21,7	21,8	21,9	22,9	17,76
10-May-14-10:38:50	23	22,1	22,3	23	18,04
10-May-14-11:08:50	25	23,8	23,4	24,5	18,84
10-May-14-11:38:50	26,1	24,4	24	24,8	19,88
10-May-14-12:08:50	26,6	25,3	24,5	24,9	20,5
10-May-14-12:38:50	28,8	25,9	25,2	25,4	20,96
10-May-14-13:08:50	24,9	23,9	24,4	24,4	21,33
10-May-14-13:38:50	25,3	24,7	25,1	25,1	21,02
10-May-14-14:08:50	26,7	25	25	25	21,45
10-May-14-14:38:50	27,3	26,1	26,1	26,2	21,96
10-May-14-15:08:50	24,6	24,2	24,8	24,9	22,12
10-May-14-15:38:50	24,1	24,2	25,1	25,3	21,59
10-May-14-16:08:50	25,2	24,9	25,4	25,8	21,79
10-May-14-16:38:50	24,3	24,6	25,2	25,9	22,1
10-May-14-17:08:50	22,7	23,3	25,1	25,1	21,61
10-May-14-17:38:50	19,2	20,2	23,1	21,9	20,78

Table 2: Data obtained between 8 and 14 May 2014 from both rooms after the green roof was installed. T1 is the temperature in front of the roof tile, T2 the air temperature in the centre of the room and Te the outside temperature.

	Green Roof		Fibre cement roof		
Date and time	T1 °C	T2 °C	T1 °C	T2 °C	Te
10-May-14-18:08:50	17,2	18,2	21,5	19,9	18,99
10-May-14-18:38:50	15,6	16,8	20,2	18,8	17,55
10-May-14-19:08:50	14,5	15,9	19,2	17,8	16,56
10-May-14-19:38:50	14,6	15,5	18,6	17,2	15,68
10-May-14-20:08:50	14,5	15,6	18,1	16,7	15,52
10-May-14-20:38:50	15,3	16,1	17,8	17	15,66
10-May-14-21:08:50	15,6	16,1	17,7	17,1	16,37
10-May-14-21:38:50	15,6	16,2	17,6	17,1	16,8
10-May-14-22:08:50	15,2	15,9	17,4	16,8	16,75
10-May-14-22:38:50	15,5	16	17,2	16,9	16,21
10-May-14-23:08:50	15,7	15,9	17,1	16,8	16,32
10-May-14-23:38:50	15	15,7	16,9	16,6	16,26
11-May-14-00:08:50	14,9	15,4	16,7	16,3	15,9
11-May-14-00:38:50	14,6	15	16,4	15,9	15,65
11-May-14-01:08:50	14,4	14,8	16,1	15,7	15,27
11-May-14-01:38:50	14,5	15	16	15,7	14,98
11-May-14-02:08:50	14,3	14,6	16	15,7	15,05
11-May-14-02:38:50	14	14,6	15,9	15,6	15,18
11-May-14-03:08:50	13,9	14,5	15,8	15,6	15,07
11-May-14-03:38:50	13,7	14,4	15,4	15,2	14,91
11-May-14-04:08:50	13,8	14,4	15,4	15,3	14,78
11-May-14-04:38:50	13,9	14,2	15,6	15,3	14,78
11-May-14-05:08:50	13,8	14,3	15,5	15,4	14,86
11-May-14-05:38:50	13,4	14	15,3	15,1	14,79
11-May-14-06:08:50	13,2	13,8	15,1	14,8	14,54
11-May-14-06:38:50	13,4	14,1	15,2	14,9	14,37
11-May-14-07:08:50	14,6	15,1	15,8	16,3	14,51
11-May-14-07:38:50	16,6	17,6	17,6	18,5	15,07
11-May-14-08:08:50	18,6	19,4	19	20,6	16,1
11-May-14-08:38:50	20	20,8	20,2	21,8	17
11-May-14-09:08:50	20,8	21,4	20,9	22,1	17,69
11-May-14-09:38:50	21,6	21,7	21,6	22,7	18,19
11-May-14-10:08:50	23,6	23,5	22,5	23,7	18,82
11-May-14-10:38:50	24	23,9	23,1	24,2	19,72

Table 2: Data obtained between 8 and 14 May 2014 from both rooms after the green roof was installed. Where T1 is the temperature in front of the roof tile, T2 is the air temperature in the centre of the room and Te is the outside temperature.(continued)

	Green Roof		Fibre cement roof		
Date and time	T1 °C	T2 °C	T1 °C	T2 °C	Te
11-May-14-11:08:50	25,3	24,4	23,5	24,5	20,37
11-May-14-11:38:50	25,7	25,2	24	24,6	20,83
11-May-14-12:08:50	27,2	25,9	24,4	25,3	21,4
11-May-14-12:38:50	27,5	26,5	24,7	25,3	21,77
11-May-14-13:08:50	28,4	26,1	25,1	25,8	21,99
11-May-14-13:38:50	28,4	27,2	25,2	25,7	22,16
11-May-14-14:08:50	27,4	26,7	25,6	26,5	22,66
11-May-14-14:38:50	30,4	28,4	26,9	28,1	22,95
11-May-14-15:08:50	27	26,9	26,4	26,7	23,99
11-May-14-15:38:50	26,6	26,4	26	26,6	23,08

11-May-14-16:08:50	26,3	26,2	26,4	27,1	22,95
11-May-14-16:38:50	24,2	25,4	26	26,2	23,07
11-May-14-17:08:50	22,2	23,2	24,9	24,8	22,31
11-May-14-17:38:50	19,9	21,2	23,8	22,9	21,26
11-May-14-18:08:50	18,2	19,1	22,5	20,8	20,15
11-May-14-18:38:50	17,4	18,5	21,5	20,2	19,13
11-May-14-19:08:50	16,5	17,8	20,8	19,6	18,87
11-May-14-19:38:50	15,8	17,2	20,2	18,7	18,22
11-May-14-20:08:50	15,5	16,4	19,8	18,2	17,69
11-May-14-20:38:50	15,3	16,3	19,2	17,8	17,36
11-May-14-21:08:50	14,6	15,6	18,7	17,3	17,43
11-May-14-21:38:50	14,3	15,6	18,2	17	17,03
11-May-14-22:08:50	14,1	15,1	17,8	16,8	16,77
11-May-14-22:38:50	14,3	15	17,6	16,6	16,47
11-May-14-23:08:50	14,9	15,6	17,6	17	16,36
11-May-14-23:38:50	15,2	15,9	17,6	17,1	16,77
12-May-14-00:08:50	15	15,8	17,5	17	16,73
12-May-14-00:38:50	14,8	15,4	17,2	16,6	16,28
12-May-14-01:08:50	14,5	15,1	16,9	16,2	15,77
12-May-14-01:38:50	13,8	14,6	16,6	15,5	15,32
12-May-14-02:08:50	13,4	14,1	16	15,2	14,71
12-May-14-02:38:50	13,1	13,9	15,8	14,9	14,29
12-May-14-03:08:50	13	13,7	15,4	14,6	14,14
12-May-14-03:38:50	12,3	13,1	15,1	14,2	13,84

Table 2: Data obtained between 8 and 14 May 2014 from both rooms after the green roof was installed. Where T1 is the temperature in front of the roof tile, T2 is the air temperature in the centre of the room and Te is the outside temperature.(continued)

	Green Roof		Fibre cement roof		
Date and time	T1 °C	T2 °C	T1 °C	T2 °C	Te
12-May-14-04:08:50	12,1	12,9	14,8	14	13,48
12-May-14-04:38:50	11,7	12,4	14,5	13,8	13,14
12-May-14-05:08:50	10,9	12,1	14	13,3	12,59
12-May-14-05:38:50	11,3	11,9	13,8	12,9	12,38
12-May-14-06:08:50	11,4	12,1	13,9	13,1	12,2
12-May-14-06:38:50	10,6	11,7	13,7	12,8	12,56
12-May-14-07:08:50	11,3	12	13,6	13,5	12,25
12-May-14-07:38:50	14,2	14,2	15,3	16,2	12,56
12-May-14-08:08:50	16,3	16,9	17,9	19,3	14,04
12-May-14-08:38:50	18,9	19,9	20,1	22	15,5
12-May-14-09:08:50	21,4	22,2	21,7	23,4	16,69
12-May-14-09:38:50	23,4	23,6	22,7	24,1	17,99
12-May-14-10:08:50	25,1	25	23,3	24,9	19,23
12-May-14-10:38:50	25,9	25,5	24,1	25,3	19,8
12-May-14-11:08:50	27,6	25,9	24,9	26	20,4
12-May-14-11:38:50	28,3	27,4	25,4	26,6	20,89
12-May-14-12:08:50	28,6	27,5	25,9	26,7	21,58
12-May-14-12:38:50	29,7	28,2	26,3	27,3	22,24
12-May-14-13:08:50	30,9	28,7	26,8	27,7	22,7
12-May-14-13:38:50	30,2	28,6	27,1	27,6	22,91
12-May-14-14:08:50	30	28,5	27,7	28,1	23,21
12-May-14-14:38:50	29,9	28,7	28,1	28,6	23,6
12-May-14-15:08:50	30,5	29,6	28,6	29,5	23,84
12-May-14-15:38:50	29,1	28,5	28,7	29,3	24,32
12-May-14-16:08:50	27,9	28,6	28,4	29,1	23,97
12-May-14-16:38:50	25,3	25,7	27,3	27,5	23,57
12-May-14-17:08:50	22	24,7	26	26,3	22,95
12-May-14-17:38:50	19,5	22,2	24,7	23,8	22

12-May-14-18:08:50	18	20,2	23,6	22,4	20,99
12-May-14-18:38:50	16,7	19,2	22,5	21,3	19,92
12-May-14-19:08:50	16	18,3	21,8	20,2	19,21
12-May-14-19:38:50	14,8	17,2	20,9	18,9	18,16
12-May-14-20:08:50	14	16,3	20	18,1	16,77
12-May-14-20:38:50	13,2	15,3	19,2	17,4	16,17

Table 2: Data obtained between 8 and 14 May 2014 from both rooms after the green roof was installed. Where T1 is the temperature in front of the roof tile, T2 is the air temperature in the centre of the room and Te is the outside temperature.(continued)

	Green Roof		Fibre cement roof		
Date and time	T1 °C	T2 °C	T1 °C	T2 °C	Te
12-May-14-21:08:50	13,4	14,9	18,6	16,7	15,26
12-May-14-21:38:50	13	14,7	18	16,4	14,57
12-May-14-22:08:50	12,2	14,1	17,6	15,7	14,66
12-May-14-22:38:50	12,3	14	17,2	15,5	14,11
12-May-14-23:08:50	12,3	13,8	16,8	15,2	14,07
12-May-14-23:38:50	12,1	13,5	16,6	15	13,99
13-May-14-00:08:50	11,6	13,3	16	14,8	13,77
13-May-14-00:38:50	10,9	12,9	15,8	14,5	13,45
13-May-14-01:08:50	11,3	12,9	15,6	14,5	13,18
13-May-14-01:38:50	11,8	12,9	15,6	14,5	13,09
13-May-14-02:08:50	10,5	12,4	15	13,8	12,89
13-May-14-02:38:50	10,7	12,4	14,8	13,9	12,42
13-May-14-03:08:50	11,7	12,8	14,8	14	13,15
13-May-14-03:38:50	11,5	13	14,9	14,1	13,33
13-May-14-04:08:50	10,5	12,3	14,6	13,7	13,35
13-May-14-04:38:50	11,4	12,9	14,7	14,1	12,92
13-May-14-05:08:50	11,5	12,8	14,6	14,2	13,41
13-May-14-05:38:50	10,9	12,3	14,5	13,8	12,85
13-May-14-06:08:50	11,9	13	14,6	14,3	12,96
13-May-14-06:38:50	10,4	12,1	14,2	13,6	13,62
13-May-14-07:08:50	12,4	13,2	14,3	14,3	12,64
13-May-14-07:38:50	12,8	13,8	14,8	15,2	13,29
13-May-14-08:08:50	17,7	18,5	17,9	19,5	14,05
13-May-14-08:38:50	18,9	20,8	20,2	21,6	17,47
13-May-14-09:08:50	22,3	22,7	21,7	23,3	18,52
13-May-14-09:38:50	22,5	22,6	21,5	22,8	19,52
13-May-14-10:08:50	25,8	24,6	22,2	24,2	20,03
13-May-14-10:38:50	26,1	26,1	24,2	25,6	21,2
13-May-14-11:08:50	28,6	28,3	25,7	27,3	21,99
13-May-14-11:38:50	28,9	27,9	25,6	27,2	22,71
13-May-14-12:08:50	31,7	29,4	26,2	28,1	22,72
13-May-14-12:38:50	30,4	29	26,9	28,5	23,72
13-May-14-13:08:50	29,6	27,6	26,6	27,4	24
13-May-14-13:38:50	31,1	28,7	26,6	27,8	24,04

Table 2: Data obtained between 8 and 14 May 2014 from both rooms after the green roof was installed. Where T1 is the temperature in front of the roof tile, T2 is the air temperature in the centre of the room and Te is the outside temperature.(continued)

	Green Roof		Fibre cement roof		
Date and time	T1 °C	T2 °C	T1 °C	T2 °C	Te
13-May-14-14:08:50	33,8	31,6	28,7	30,2	24,36
13-May-14-14:38:50	32,2	31,4	29,3	30,5	25,48
13-May-14-15:08:50	30,5	30,1	28,6	29,7	25,06
13-May-14-15:38:50	28,1	28,5	28,2	28,6	24,92
13-May-14-16:08:50	25,7	26,4	26,9	26,9	24,25

13-May-14-16:38:50	24,9	25,6	26,2	26,2	23,65
13-May-14-17:08:50	23,7	24,2	25,8	25,4	23,56
13-May-14-17:38:50	22	22,9	25,1	24,1	23,08
13-May-14-18:08:50	20,9	21,8	24,1	23,1	21,89
13-May-14-18:38:50	19,8	20,9	23,5	22,4	21,32
13-May-14-19:08:50	19,5	20,5	22,9	22	20,43
13-May-14-19:38:50	19	19,9	22,3	21,2	20,15
13-May-14-20:08:50	18,9	19,6	22,1	21,1	19,38
13-May-14-20:38:50	18,6	19,2	21,7	20,6	19,31
13-May-14-21:08:50	18,4	19	21,4	20,5	18,72
13-May-14-21:38:50	18,3	19,1	21,1	20,3	18,65
13-May-14-22:08:50	17,9	18,7	20,9	19,9	18,84
13-May-14-22:38:50	18,2	18,7	20,8	19,8	18,74
13-May-14-23:08:50	17,8	18,6	20,5	19,8	18,87
13-May-14-23:38:50	17,7	18,4	20,1	19,4	18,52
14-May-14-00:08:50	18,1	18,4	20	19,4	18,42
14-May-14-00:38:50	18,2	18,5	19,9	19,5	18,25
14-May-14-01:08:50	18,1	18,7	20,1	19,6	18,27
14-May-14-01:38:50	18,1	18,7	19,7	19,4	18,24
14-May-14-02:08:50	17,7	18,3	19,7	19,4	18,25
14-May-14-02:38:50	17,8	18,4	19,5	19,4	18,42
14-May-14-03:08:50	18,1	18,5	19,6	19,5	18,51
14-May-14-03:38:50	17,7	18,3	19,5	19,3	18,58
14-May-14-04:08:50	17,9	18,3	19,4	19,4	18,14
14-May-14-04:38:50	18,3	18,6	19,5	19,4	18,52
14-May-14-05:08:50	18,2	18,6	19,4	19,4	18,5
14-May-14-05:38:50	18,3	18,5	19,4	19,3	18,41
14-May-14-06:08:50	18,3	18,7	19,4	19,4	18,34
14-May-14-06:38:50	17,9	18,4	19,5	19,2	18,42

Table 2: Data obtained between 8 and 14 May 2014 from both rooms after the green roof was installed. Where T1 is the temperature in front of the roof tile, T2 is the air temperature in the centre of the room and Te is the outside temperature.(conclusion)

	Green Roof		Fibre cement roof		
Date and time	T1 °C	T2 °C	T1 °C	T2 °C	Te
14-May-14-07:08:50	18	18,5	19,5	19,3	17,81
14-May-14-07:38:50	18,6	18,8	19,4	19,4	18,09
14-May-14-08:08:50	18,9	19,2	19,7	20	18,38
14-May-14-08:38:50	18,8	19,1	20	19,9	18,6
14-May-14-09:08:50	19,2	19,2	19,9	19,8	18,22
14-May-14-09:38:50	19,7	19,6	20,4	20,2	18,12
14-May-14-10:08:50	19,4	19,6	20,4	20,5	18,43
14-May-14-10:38:50	19,4	19,5	20,3	20,2	18,43
14-May-14-11:08:50	19,2	19,2	20	19,9	18,31
14-May-14-11:38:50	19,7	19,7	20	20,3	18
14-May-14-12:08:50	19,6	19,7	20,1	20,3	18,27
14-May-14-12:38:50	19,6	19,6	20	20,2	18,37
14-May-14-13:08:50	19,1	19,3	19,8	19,9	18,32
14-May-14-13:38:50	20,2	20	20,1	20,6	18,18
14-May-14-14:08:50	19	19,2	19,9	19,7	18,33
14-May-14-14:38:50	19,4	19,5	20,1	20	18,07
14-May-14-15:08:50	20	19,6	20	20,1	18,14
14-May-14-15:38:50	19,9	19,8	20,1	20,3	18,28
14-May-14-16:08:50	19,6	19,6	20,1	20,1	18,33
14-May-14-16:38:50	19,2	19,5	19,9	20	18,35
14-May-14-17:08:50	19,2	19,4	20,2	20,4	18,27
14-May-14-17:38:50	17,8	18,4	19,7	19,3	18,14
14-May-14-18:08:50	17,1	17,7	19	18,6	17,41

14-May-14-18:38:50	16,5	17,2	18,4	18	17
14-May-14-19:08:50	16,1	16,8	18,2	17,8	16,71
14-May-14-19:38:50	16,2	16,7	18,3	17,7	16,54
14-May-14-20:08:50	15,2	16	17,8	17,1	16,66
14-May-14-20:38:50	16	16,4	17,8	17,1	16,57
14-May-14-21:08:50	15,9	16,3	17,5	17,1	16,77
14-May-14-21:38:50	15,7	16,4	17,4	17,2	15,97
14-May-14-22:08:50	15,7	16,3	17,3	17,3	16,12
14-May-14-22:38:50	15,4	16,1	17,2	17,1	16,37
14-May-14-23:08:50	15,8	16,2	17,3	17,2	16,09
14-May-14-23:38:50	14,9	15,8	17	16,8	15,93

Table 3: Data obtained between 2 and 7 May 2014 for T1-T2 (roof temperature minus internal air temperature) and solar radiation values provided by the campus weather station. (continued)

date and time	fibre cement roof	green roof	solar radiation (W/m2)
02-May-14-00:00	-1,4	1,2	0
02-May-14-00:30	-1,5	1,5	0
02-May-14-01:00	-1,4	1,2	0
02-May-14-01:30	-1,2	1	0
02-May-14-02:00	-1	1	0
02-May-14-02:30	-1,4	1,2	0
02-May-14-03:00	-0,7	1,1	0
02-May-14-03:30	-0,6	0,6	0
02-May-14-04:00	-0,7	0,4	0
02-May-14-04:30	-1	0,5	0
02-May-14-05:00	-0,3	0	0
02-May-14-05:30	-1,2	0,4	0
02-May-14-06:00	-1,2	0,5	0
02-May-14-06:30	0,2	0,4	0
02-May-14-07:00	0,6	-0,2	0,5
02-May-14-07:30	0,5	-0,5	18,31
02-May-14-08:00	0,3	-0,7	42,81
02-May-14-08:30	0,7	-0,9	62,44
02-May-14-09:00	0,5	-1,2	133,5
02-May-14-09:30	-0,3	-1,9	272,69
02-May-14-10:00	0	-1,9	413,12
02-May-14-10:30	0,4	-2,1	478,69
02-May-14-11:00	1,1	-2	577,06
02-May-14-11:30	1,1	-1,5	193,06
02-May-14-12:00	2,5	-1,3	84,12
02-May-14-12:30	2,2	-1,5	107,06
02-May-14-13:00	2,2	-1,3	207,44
02-May-14-13:30	2,2	-1,1	610,81
02-May-14-14:00	2,8	-0,8	523,56
02-May-14-14:30	2,1	-1,1	99,56
02-May-14-15:00	1,1	-1,4	84
02-May-14-15:30	1,1	-1,2	120,56
02-May-14-16:00	-0,5	-1,1	86,19
02-May-14-16:30	-1,1	-0,1	68,75
02-May-14-17:00	-1,4	0,1	69,12

Table 3. Data obtained during the 2nd and 7th of May 2014 from T1-T2 (roof temperature minus indoor air temperature) and solar radiation values provided by the campus weather station.(continued)

Dara and time	fibre cement roof	green roof	solar radiation (W/m2)
02-May-14-00:00	-1,4	1,2	0
02-May-14-00:30	-1,5	1,5	0
02-May-14-01:00	-1,4	1,2	0
02-May-14-01:30	-1,2	1	0
02-May-14-02:00	-1	1	0
02-May-14-02:30	-1,4	1,2	0
02-May-14-03:00	-0,7	1,1	0
02-May-14-03:30	-0,6	0,6	0
02-May-14-04:00	-0,7	0,4	0
02-May-14-04:30	-1	0,5	0
02-May-14-05:00	-0,3	0	0
02-May-14-05:30	-1,2	0,4	0
02-May-14-06:00	-1,2	0,5	0
02-May-14-06:30	0,2	0,4	0
02-May-14-07:00	0,6	-0,2	0,5
02-May-14-07:30	0,5	-0,5	18,31
02-May-14-08:00	0,3	-0,7	42,81
02-May-14-08:30	0,7	-0,9	62,44
02-May-14-09:00	0,5	-1,2	133,5
02-May-14-09:30	-0,3	-1,9	272,69
02-May-14-10:00	0	-1,9	413,12
02-May-14-10:30	0,4	-2,1	478,69
02-May-14-11:00	1,1	-2	577,06
02-May-14-11:30	1,1	-1,5	193,06
02-May-14-12:00	2,5	-1,3	84,12
02-May-14-12:30	2,2	-1,5	107,06
02-May-14-13:00	2,2	-1,3	207,44
02-May-14-13:30	2,2	-1,1	610,81
02-May-14-14:00	2,8	-0,8	523,56
02-May-14-14:30	2,1	-1,1	99,56
02-May-14-15:00	1,1	-1,4	84
02-May-14-15:30	1,1	-1,2	120,56
02-May-14-16:00	-0,5	-1,1	86,19
02-May-14-16:30	-1,1	-0,1	68,75
02-May-14-17:00	-1,4	0,1	69,12

Table 3: Data obtained on 2 and 7 May 2014 for T1-T2 (roof temperature minus indoor air temperature) and solar radiation values provided by the campus weather station.(continued)

date and time	fibre cement roof	green roof	solar radiation (W/m2)
02-May-14-00:00	-1,4	1,2	0
02-May-14-00:30	-1,5	1,5	0
02-May-14-01:00	-1,4	1,2	0
02-May-14-01:30	-1,2	1	0
02-May-14-02:00	-1	1	0
02-May-14-02:30	-1,4	1,2	0
02-May-14-03:00	-0,7	1,1	0
02-May-14-03:30	-0,6	0,6	0
02-May-14-04:00	-0,7	0,4	0
02-May-14-04:30	-1	0,5	0
02-May-14-05:00	-0,3	0	0
02-May-14-05:30	-1,2	0,4	0
02-May-14-06:00	-1,2	0,5	0
02-May-14-06:30	0,2	0,4	0
02-May-14-07:00	0,6	-0,2	0,5
02-May-14-07:30	0,5	-0,5	18,31
02-May-14-08:00	0,3	-0,7	42,81

02-May-14-08:30	0,7	-0,9	62,44
02-May-14-09:00	0,5	-1,2	133,5
02-May-14-09:30	-0,3	-1,9	272,69
02-May-14-10:00	0	-1,9	413,12
02-May-14-10:30	0,4	-2,1	478,69
02-May-14-11:00	1,1	-2	577,06
02-May-14-11:30	1,1	-1,5	193,06
02-May-14-12:00	2,5	-1,3	84,12
02-May-14-12:30	2,2	-1,5	107,06
02-May-14-13:00	2,2	-1,3	207,44
02-May-14-13:30	2,2	-1,1	610,81
02-May-14-14:00	2,8	-0,8	523,56
02-May-14-14:30	2,1	-1,1	99,56
02-May-14-15:00	1,1	-1,4	84
02-May-14-15:30	1,1	-1,2	120,56
02-May-14-16:00	-0,5	-1,1	86,19
02-May-14-16:30	-1,1	-0,1	68,75
02-May-14-17:00	-1,4	0,1	69,12

Table 3: Data obtained during the 2nd and 7th of May 2014 for T1-T2 (roof temperature minus indoor air temperature) and solar radiation values provided by the campus weather station.(continued)

Date and time	fibre cement roof	green roof	solar radiation (W/m2)
02-May-14-17:00	-1,4	0,1	69,12
02-May-14-17:30	-0,9	0,9	72,25
02-May-14-18:00	-1,8	1,6	7,19
02-May-14-18:30	-1,4	2	0,25
02-May-14-19:00	-1,7	1,9	0
02-May-14-19:30	-1,2	1,6	0
02-May-14-20:00	-1,4	1,7	0
02-May-14-20:30	-1,2	1,9	0
02-May-14-21:00	-1,5	1,9	0
02-May-14-21:30	-0,7	1,4	0
02-May-14-22:00	-1,2	1,5	0
02-May-14-22:30	-1,4	1,6	0
02-May-14-23:00	-1,4	1,3	0
02-May-14-23:30	-1,3	1,3	0
03-May-14-00:00	-1	1,4	0
03-May-14-00:30	-0,7	1	0
03-May-14-01:00	-1,1	0,8	0
03-May-14-01:30	-1,6	0,9	0
03-May-14-02:00	-0,9	0,8	0
03-May-14-02:30	-0,7	0,9	0
03-May-14-03:00	-0,6	0,6	0
03-May-14-03:30	-1,2	0,7	0
03-May-14-04:00	-0,6	0,2	0
03-May-14-04:30	-1	0,4	0
03-May-14-05:00	-0,2	0,5	0
03-May-14-05:30	-0,6	0,4	0
03-May-14-06:00	-0,4	0,5	0
03-May-14-06:30	-0,5	0,4	0
03-May-14-07:00	-0,4	-0,4	1,62
03-May-14-07:30	-0,6	-1,2	40,88
03-May-14-08:00	-0,4	-2,2	131,94
03-May-14-08:30	-1,3	-2,1	234,75
03-May-14-09:00	-0,9	-2,1	330,38
03-May-14-09:30	-0,7	-1,7	444,12
03-May-14-10:00	0	-2	534,69

03-May-14-10:30	0,4	-1,4	608,25

Table 3: Data obtained during the 2nd and 7th of May 2014 for T1-T2 (roof temperature minus indoor air temperature) and solar radiation values provided by the campus weather station.(continued)

Date and time	fibre cement roof	green roof	solar radiation (W/m2)
03-May-14-10:30	0,4	-1,4	608,25
03-May-14-11:00	0,5	-1,2	667,25
03-May-14-11:30	1,6	-1,2	151,5
03-May-14-12:00	2,1	-1,2	45,31
03-May-14-12:30	2,6	-1	42,56
03-May-14-13:00	1,3	-0,9	152
03-May-14-13:30	0,8	-1	711,62
03-May-14-14:00	0,8	-0,8	654,19
03-May-14-14:30	1,3	-1,5	45,19
03-May-14-15:00	0,2	-1,7	39,06
03-May-14-15:30	0	-1,4	55,19
03-May-14-16:00	-0,4	-1,7	60,81
03-May-14-16:30	-0,6	-1	37,94
03-May-14-17:00	-1	-0,6	31,62
03-May-14-17:30	-1,4	0,6	58
03-May-14-18:00	-1,6	1	7,69
03-May-14-18:30	-1,5	1,4	0,12
03-May-14-19:00	-1,6	1,5	0
03-May-14-19:30	-1,5	1,7	0
03-May-14-20:00	-1,2	1,6	0
03-May-14-20:30	-1,2	1,6	0
03-May-14-21:00	-1,4	1,9	0
03-May-14-21:30	-1,1	1,8	0
03-May-14-22:00	-0,9	1,2	0
03-May-14-22:30	-1,6	1,4	0
03-May-14-23:00	-1	1	0
03-May-14-23:30	-0,7	0,8	0
04-May-14-00:00	-0,3	1	0
04-May-14-00:30	-0,3	0,7	0
04-May-14-01:00	-0,8	0,7	0
04-May-14-01:30	-0,6	0,5	0
04-May-14-02:00	-1,3	0,6	0
04-May-14-02:30	-1,4	0,9	0
04-May-14-03:00	-0,9	0,9	0
04-May-14-03:30	-0,7	0,6	0
04-May-14-04:00	-1,2	1	0

Table 3: Data obtained during the 2nd and 7th of May 2014 for T1-T2 (roof temperature minus indoor air temperature) and solar radiation values provided by the campus weather station.(continued)

Date and time	fibre cement roof	green roof	solar radiation (W/m2)
04-May-14-04:00	-1,2	1	0
04-May-14-04:30	-0,7	1	0
04-May-14-05:00	-0,8	1,1	0
04-May-14-05:30	-1	1,2	0
04-May-14-06:00	-0,8	0,9	0

04-May-14-06:30	-1,2	0,9	0
04-May-14-07:00	-0,3	-0,1	3,62
04-May-14-07:30	0,2	-0,8	49,25
04-May-14-08:00	-0,2	-1,3	111,44
04-May-14-08:30	-0,9	-1,6	168,25
04-May-14-09:00	-0,6	-1,6	246,06
04-May-14-09:30	-0,4	-1,7	349,94
04-May-14-10:00	0	-1,9	494
04-May-14-10:30	0	-1,6	571,62
04-May-14-11:00	0	-1,4	543,44
04-May-14-11:30	0,9	-1,3	194,62
04-May-14-12:00	0,3	-1,3	182,81
04-May-14-12:30	0,1	-1,1	160,88
04-May-14-13:00	1,1	-1,2	236
04-May-14-13:30	0,7	-1,1	353,31
04-May-14-14:00	1,1	-1,3	273,69
04-May-14-14:30	0	-0,9	190,69
04-May-14-15:00	0	-0,8	213,94
04-May-14-15:30	0,2	-1,2	205,69
04-May-14-16:00	-0,2	-1,3	200,12
04-May-14-16:30	-1	-0,7	137,19
04-May-14-17:00	-1,1	0	69,94
04-May-14-17:30	-0,8	0,6	44,38
04-May-14-18:00	-1,7	1	7
04-May-14-18:30	-1,7	1,3	0
04-May-14-19:00	-1,2	1,4	0
04-May-14-19:30	-1,4	1,5	0
04-May-14-20:00	-1,2	1,2	0
04-May-14-20:30	-0,9	1	0
04-May-14-21:00	-0,8	0,8	0
04-May-14-21:30	-1,3	0,9	0

Table 3: Data obtained during the 2nd and 7th of May 2014 for T1-T2 (roof temperature minus indoor air temperature) and solar radiation values provided by the campus weather station.(continued)

Date and time	fibre cement roof	green roof	solar radiation (W/m2)
04-May-14-21:30	-1,3	0,9	0
04-May-14-22:00	-0,6	0,8	0
04-May-14-22:30	-1,2	1,1	0
04-May-14-23:00	-0,5	0,6	0
04-May-14-23:30	-0,5	0,5	0
05-May-14-00:00	-0,6	0,8	0
05-May-14-00:30	-0,6	0,6	0
05-May-14-01:00	-0,6	0,8	0
05-May-14-01:30	-0,8	0,3	0
05-May-14-02:00	-0,6	0,3	0
05-May-14-02:30	-0,9	0,5	0
05-May-14-03:00	-0,7	0,5	0
05-May-14-03:30	-0,8	0,5	0
05-May-14-04:00	-0,7	0,5	0
05-May-14-04:30	-0,8	0,3	0
05-May-14-05:00	-1	0,6	0
05-May-14-05:30	-0,7	0,3	0
05-May-14-06:00	-0,5	0,4	0
05-May-14-06:30	-0,6	0,3	0
05-May-14-07:00	-0,4	0,3	0,12
05-May-14-07:30	0,3	-0,2	3,56
05-May-14-08:00	-0,2	-0,4	31,94
05-May-14-08:30	-0,1	-0,4	70,44
05-May-14-09:00	0,4	-0,9	85,88
05-May-14-09:30	0,3	-0,9	122,88

05-May-14-10:00	0,3	-1	100,5
05-May-14-10:30	0,7	-1	97,56
05-May-14-11:00	0,5	-1	144,25
05-May-14-11:30	0,2	-0,8	138,62
05-May-14-12:00	0	-0,7	109,94
05-May-14-12:30	1,6	-1	131,69
05-May-14-13:00	1,5	-1,1	257,38
05-May-14-13:30	0,7	-1	233,06
05-May-14-14:00	0,9	-1,2	192,69
05-May-14-14:30	0,4	-1	219,62
05-May-14-15:00	0,5	-0,6	209,69

Table 3: Data obtained during the 2nd and 7th of May 2014 for T1-T2 (roof temperature minus indoor air temperature) and solar radiation values provided by the campus weather station.(continued)

Date and time	fibre cement roof	green roof	solar radiation (W/m2)
05-May-14-15:00	0,5	-0,6	209,69
05-May-14-15:30	0,2	-0,5	182,69
05-May-14-16:00	-0,5	-0,4	149,31
05-May-14-16:30	-0,5	-0,1	100,5
05-May-14-17:00	-0,6	0,2	57,19
05-May-14-17:30	-0,7	0,4	22
05-May-14-18:00	-0,7	0,6	3,25
05-May-14-18:30	-0,5	0,6	0
05-May-14-19:00	-0,6	0,7	0
05-May-14-19:30	-0,8	0,5	0
05-May-14-20:00	-0,9	0,6	0
05-May-14-20:30	-0,7	0,5	0
05-May-14-21:00	-0,5	0,7	0
05-May-14-21:30	-0,8	0,4	0
05-May-14-22:00	-0,6	0,6	0
05-May-14-22:30	-0,5	0,4	0
05-May-14-23:00	-0,7	0,5	0
05-May-14-23:30	-0,7	0,6	0
06-May-14-00:00	-0,4	0,5	0
06-May-14-00:30	-0,5	0,5	0
06-May-14-01:00	-0,5	0,4	0
06-May-14-01:30	-0,7	0,3	0
06-May-14-02:00	-0,7	0,6	0
06-May-14-02:30	-0,8	0,4	0
06-May-14-03:00	-0,9	0,7	0
06-May-14-03:30	-1,1	0,6	0
06-May-14-04:00	-0,9	0,3	0
06-May-14-04:30	-0,7	0,4	0
06-May-14-05:00	-0,7	0,4	0
06-May-14-05:30	-0,4	0,2	0
06-May-14-06:00	-0,2	0	0
06-May-14-06:30	-0,7	0,1	0
06-May-14-07:00	-0,3	0,1	0,38
06-May-14-07:30	-0,3	-0,2	16,69
06-May-14-08:00	-0,2	-0,7	56,94
06-May-14-08:30	-0,1	-1	108,88

Table 3: Data obtained during the 2nd and 7th of May 2014 for T1-T2 (roof temperature minus

indoor air temperature) and solar radiation values provided by the campus weather station.(continued)

Date and time	fibre cement roof	green roof	solar radiation (W/m2)
06-May-14-08:30	-0,1	-1	108,88
06-May-14-09:00	0	-1,4	182,81
06-May-14-09:30	0,2	-1,5	239,38
06-May-14-10:00	0,4	-1,5	303,19
06-May-14-10:30	0,5	-1,4	417,06
06-May-14-11:00	0,4	-1,4	293,75
06-May-14-11:30	0,2	-1,1	274,44
06-May-14-12:00	2,1	-1,3	239,75
06-May-14-12:30	3	-1,5	188,38
06-May-14-13:00	2,2	-1,2	256,12
06-May-14-13:30	2,1	-1,2	454,25
06-May-14-14:00	1,3	-1	479,31
06-May-14-14:30	1,1	-1,1	180,81
06-May-14-15:00	1,2	-1,3	126,62
06-May-14-15:30	0,4	-1,1	120,44
06-May-14-16:00	-1,2	-0,9	60,75
06-May-14-16:30	-1,3	-0,4	61,88
06-May-14-17:00	-1,8	0,1	74,31
06-May-14-17:30	-1,9	0,5	48,19
06-May-14-18:00	-2	1,1	7,5
06-May-14-18:30	-2	1,2	0
06-May-14-19:00	-2	1	0
06-May-14-19:30	-2,1	1,1	0
06-May-14-20:00	-1,7	1,2	0
06-May-14-20:30	-1,6	1	0
06-May-14-21:00	-1,7	1,2	0
06-May-14-21:30	-2	1,3	0
06-May-14-22:00	-1,7	1,5	0
06-May-14-22:30	-1,9	1,3	0
06-May-14-23:00	-1,6	1,2	0
06-May-14-23:30	-1,6	1,1	0
07-May-14-00:00	-1,5	1,1	0
07-May-14-00:30	-1,4	1,1	0
07-May-14-01:00	-1,5	0,8	0
07-May-14-01:30	-1,3	0,9	0
07-May-14-02:00	-1,6	1	0

Table 3: Data obtained during the 2nd and 7th of May 2014 for T1-T2 (roof temperature minus indoor air temperature) and solar radiation values provided by the campus weather

station.(continued)

Date and time	fibre cement roof	green roof	solar radiation (W/m2)
07-May-14-02:00	-1,6	1	0
07-May-14-02:30	-1,5	1	0
07-May-14-03:00	-1,7	0,9	0
07-May-14-03:30	-1,2	0,5	0
07-May-14-04:00	-2,1	0,8	0
07-May-14-04:30	-1,2	0,6	0
07-May-14-05:00	-1,3	0,6	0
07-May-14-05:30	-1,7	0,7	0
07-May-14-06:00	-2	0,7	0
07-May-14-06:30	-1,6	0,8	0
07-May-14-07:00	-1,3	-0,1	0,38
07-May-14-07:30	-1,6	-1,2	31,62
07-May-14-08:00	-2,9	-1,9	122,38
07-May-14-08:30	-2,8	-2,4	233,69
07-May-14-09:00	-2,9	-2,6	329,75
07-May-14-09:30	-1,6	-2,1	418
07-May-14-10:00	0	-2,1	492,62
07-May-14-10:30	1,1	-1,8	537,12
07-May-14-11:00	2,1	-2	623,94
07-May-14-11:30	3	-1,6	182,81
07-May-14-12:00	3,8	-1,5	48,5
07-May-14-12:30	3,8	-0,7	48,12
07-May-14-13:00	4,5	-1,1	142,94
07-May-14-13:30	4,4	-1,4	688,06
07-May-14-14:00	2,9	-1,1	637,62
07-May-14-14:30	2,4	-1,3	64,5
07-May-14-15:00	1,5	-1,1	36,69
07-May-14-15:30	0,5	-0,9	36,38
07-May-14-16:00	-0,6	-0,8	35,12
07-May-14-16:30	-1,8	-0,8	31,81
07-May-14-17:00	-2,6	-0,3	27,56
07-May-14-17:30	-3	1,2	49,75
07-May-14-18:00	-2,7	1,9	5,62
07-May-14-18:30	-2,3	1,9	0,12
07-May-14-19:00	-2,2	2,1	0
07-May-14-19:30	-2,4	2	0

Table 3: Data obtained during the 2nd and 7th of May 2014 from T1-T2 (roof temperature minus indoor air temperature) and solar radiation values provided by the campus weather station.(conclusion)

Date and time	fibre cement roof	green roof	solar radiation (W/m2)
07-May-14-19:30	-2,4	2	0
07-May-14-20:00	-2,4	2,3	0
07-May-14-20:30	-2,2	2,1	0
07-May-14-21:00	-2,1	1,9	0
07-May-14-21:30	-1,8	1,9	0
07-May-14-22:00	-2,2	1,9	0
07-May-14-22:30	-1,9	2	0
07-May-14-23:00	-2	1,7	0
07-May-14-23:30	-2,1	1,8	0

APPENDIX C

Example code used to generate the graphs in Pyhton:

```
from matplotlib.dates import strpdate2num
from matplotlib.ticker import MultipleLocator, FormatStrFormatter
import matplotlib.dates as mdates
import numpy as np
from pylab import *

datafile = ('testesolar.csv')
print 'loading', datafile

times, deltaT1, deltaT2, solarrad  = np.loadtxt(
   datafile, delimiter=',',
   converters={0:strpdate2num('%d-%b-%y-%H:%M') },
   skiprows=1, usecols=(0,1,2,3), unpack=True)

fig = figure()
ax1 = fig.add_subplot(211)
ax1.set_title('Cobertura Verde')
ax1.plot_date(times, deltaT2, 'b-',linewidth=2.0)
ax1.autoscale_view(tight='True', scalex=True, scaley=False)
ax1.grid('on')
ax1.set_ylim(-15,5)
for tl in ax1.get_yticklabels():
   tl.set_color('blue')

#
ax2 = ax1.twinx()
ax2.plot_date(times, solarrad, 'r',linewidth=2.0)
ax2.autoscale_view(tight='True', scalex=True, scaley=False)
ax2.grid('on')
ax2.set_ylim(0, 800)
for tl in ax2.get_yticklabels():
   tl.set_color('red')

locator  = MultipleLocator(1)
ax1.xaxis.set_major_locator(locator)
setp(gca(), 'xticklabels', [])

datafile = ('testesolar.csv')
print 'loading', datafile
```

```
times, deltaT1, deltaT2, solarrad = np.loadtxt(
   datafile, delimiter=',',
   converters={0:strpdate2num('%d-%b-%y-%H:%M') },
   skiprows=1, usecols=(0,1,2,3), unpack=True)

ax3 = fig.add_subplot(212)
ax3.set_title('Telhado de Fibrocimento')
ax3.plot_date(times, deltaT1, 'b-',linewidth=2.0)

ax3.set_ylabel('T1 - T2 em graus Celsius', color='b')
ax3.yaxis.set_label_coords(-0.065, 1.2)
ax3.set_ylim(-15,7)
for tl in ax3.get_yticklabels():
   tl.set_color('blue')

ax3.autoscale_view(tight='True', scalex=True, scaley=False)
ax3.grid('on')
#
ax4 = ax3.twinx()
ax4.plot_date(times, solarrad, 'r',linewidth=2.0)
ax4.set_ylabel('Fluxo solar em Watts por metro quadrado', color='r')
ax4.yaxis.set_label_coords(1.08, 1.2)
ax4.set_ylim(0, 800)
for tl in ax4.get_yticklabels():
   tl.set_color('red')

ax4.autoscale_view(tight='True', scalex=True, scaley=False)
ax4.grid('on')

# Imprime as datas no eixo x
locator  = MultipleLocator(1)
formatter = mdates.DateFormatter('%a\n%d-%b-%y')
ax3.xaxis.set_major_formatter(formatter)
ax3.xaxis.set_major_locator(locator)

#  Inclina as datas no eixo x para melhorar apresentacao
fig.autofmt_xdate()
show()
```

yes

I want morebooks!

Buy your books fast and straightforward online - at one of world's fastest growing online book stores! Environmentally sound due to Print-on-Demand technologies.

Buy your books online at
www.morebooks.shop

Kaufen Sie Ihre Bücher schnell und unkompliziert online – auf einer der am schnellsten wachsenden Buchhandelsplattformen weltweit! Dank Print-On-Demand umwelt- und ressourcenschonend produzi ert.

Bücher schneller online kaufen
www.morebooks.shop

info@omniscriptum.com
www.omniscriptum.com

Printed by Books on Demand GmbH, Norderstedt / Germany